高等院校信息技术规划教材

Spring MVC 开发技术指南

陈恒 主编

楼偶俊 巩庆志 林徐 副主编

清华大学出版社

北京

内 容 简 介

Spring MVC 是一款优秀的、基于 MVC 思想的应用框架，它是 Spring 的一个子框架。本书是一本开发技术指南，用大量的实例介绍了 Spring MVC 框架的基本思想、方法和技术，同时配备了相应的实践环节巩固 Spring MVC 应用开发的方法和技术，力图达到“做中学，学中做”。

全书共分 10 章，内容包括 Spring MVC 入门、Controller、类型转换和格式化、数据绑定和表单标签库、数据验证、国际化、文件的上传与下载、统一异常处理、EL 与 JSTL 以及名片管理系统的设计与实现等重要内容。书中实例侧重实用性和启发性，趣味性强、通俗易懂，使读者能够快速掌握 Spring MVC 框架的基础知识、编程技巧以及完整的开发体系，为适应实战应用打下坚实的基础。

本书可以作为大学计算机及相关专业的教材或教学参考书，也适合作为 Spring MVC 应用开发人员的参考用书。

本书封面贴有清华大学出版社防伪标签，无标签者不得销售。
版权所有，侵权必究。举报：010-62782989，beiqinquan@tup.tsinghua.edu.cn。

图书在版编目（CIP）数据

Spring MVC 开发技术指南/陈恒主编. —北京：清华大学出版社，2017（2024.12重印）
（高等院校信息技术规划教材）
ISBN 978-7-302-47504-0

Ⅰ. ①S… Ⅱ. ①陈… Ⅲ. ①JAVA 语言－程序设计－指南 Ⅳ. ①TP312.8－62

中国版本图书馆 CIP 数据核字（2017）第 142130 号

责任编辑：张 玥 薛 阳
封面设计：常雪影
责任校对：时翠兰
责任印制：宋 林

出版发行：清华大学出版社
网 址：https://www.tup.com.cn，https://www.wqxuetang.com
地 址：北京清华大学学研大厦 A 座 邮 编：100084
社 总 机：010-83470000 邮 购：010-62786544
投稿与读者服务：010-62776969，c-service@tup.tsinghua.edu.cn
质 量 反 馈：010-62772015，zhiliang@tup.tsinghua.edu.cn
课 件 下 载：https://www.tup.com.cn, 010-83470236
印 装 者：三河市人民印务有限公司
经 销：全国新华书店
开 本：185mm×260mm **印 张：**13.25 **字 数：**309 千字
版 次：2017 年 10 月第 1 版 **印 次：**2024 年12月第 9 次印刷
定 价：49.50 元

产品编号：075116-02

前言 Foreword

目前，尽管市面上有许多与 Spring 框架有关的书籍，但单独介绍 Spring MVC 子框架的书籍还寥寥无几。而且相关书籍非常注重知识的系统性，使得知识体系结构过于全面、庞大。这种知识体系过于庞大的书籍不太适合作为高校计算机相关专业的教材。同时，许多教师在教学过程中，非常希望教材本身能引导学生尽可能地参与到教学活动中，因此本书的重点不是简单地介绍 Spring MVC 子框架的基础知识，而是大量的实例与实践环节。读者通过本书可以快速地掌握 Spring MVC 子框架，提高 Java Web 应用的开发能力。全书共 10 章，各章的具体内容如下：

第 1 章重点讲解 MVC 的设计思想以及 Spring MVC 开发环境的构建。

第 2 章详细讲解基于注解的控制器、Controller 接收请求参数的方式以及如何编写请求处理方法，是本书的重点内容之一。

第 3 章介绍类型转换器和格式化转换器，包括内置的类型转换器和格式化转换器以及自定义类型转换器和格式化转换器。

第 4 章讲解数据绑定和表单标签库，是本书的重点内容之一。

第 5 章详细讲解 Spring MVC 框架的输入验证体系，包括 Spring 验证和 JSR303 验证，是本书的重点内容之一。

第 6 章介绍 Spring MVC 国际化的实现方法。

第 7 章讲解如何使用 Spring MVC 框架进行文件的上传与下载。

第 8 章详细讲解如何使用 Spring MVC 框架进行异常的统一处理，是本书的重点内容之一。

第 9 章介绍 EL 与 JSTL 的基本用法。

第 10 章是本书的重点内容之一，它将前面章节的知识进行综合，详细地讲解了如何使用 Spring MVC 框架来开发一个 Web 应用（名片管理系统）。

本书特别注重引导学生参与课堂教学活动，适合作为大学计算机及相关专业的教材或教学参考书，也适合作为 Spring MVC 应用开发人员的参考用书。

为便于教学，本书配有教学课件、源代码以及实践环节与课后习题参考答案，读者可从清华大学出版社网站免费下载，也可加入教材交流QQ群（46696527）下载。

由于编者水平有限，书中难免会有不足之处，敬请广大读者批评指正。

编　者
2017年2月

目录 Contents

第1章 chapter 1

Spring MVC 入门

学习目的与要求

本章重点讲解MVC的设计思想以及Spring MVC开发环境的构建。通过本章的学习，读者应了解 Spring MVC 基本流程，掌握 Spring MVC 开发环境的构建。

本章主要内容

- Spring MVC 工作原理
- Spring MVC 开发环境的构建
- 第一个 Spring MVC 应用

MVC 的中心思想将一个应用分成三个基本部分：Model（模型）、View（视图）和 Controller（控制器），让这三个部分以最低的耦合进行协同工作，从而提高应用的可扩展性及可维护性。Spring MVC 是一款优秀的基于 MVC 思想的应用框架，它是 Spring 的一个子框架。

1.1 MVC 模式与 Spring MVC 工作原理

1.1.1 MVC 模式

1. MVC 的概念

MVC 是 Model、View 和 Controller 的缩写，分别代表 Web 应用程序中的三种职责：

- 模型——用于存储数据以及处理用户请求的业务逻辑。
- 视图——向控制器提交数据，显示模型中的数据。
- 控制器——根据视图提出的请求，判断将请求和数据交给哪个模型处理，处理后的有关结果交给哪个视图更新显示。

2. 基于 Servlet 的 MVC 模式

基于 Servlet 的 MVC 模式的具体实现如下：

- 模型：一个或多个 JavaBean 对象，用于存储数据（实体模型，由 JavaBean 类创建）和处理业务逻辑（业务模型，由一般的 Java 类创建）。

- 视图：一个或多个 JSP 页面，向控制器提交数据和为模型提供数据显示，JSP 页面主要使用 HTML 标记和 JavaBean 标记来显示数据。
- 控制器：一个或多个 Servlet 对象，根据视图提交的请求进行控制，即将请求转发给处理业务逻辑的 JavaBean，并将处理结果存放到实体模型 JavaBean 中，输出给视图显示。

基于 Servlet 的 MVC 模式的流程如图 1.1 所示。

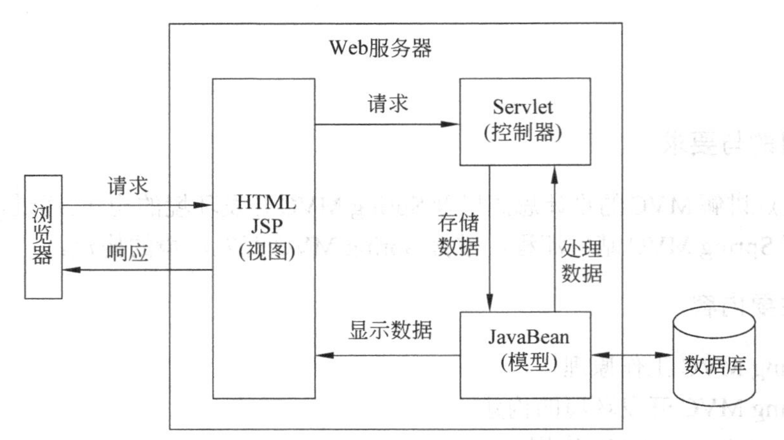

图 1.1　JSP 中的 MVC 模式

1.1.2　Spring MVC 工作原理

Spring MVC 框架是高度可配置的，包含多种视图技术，如 JSP 技术、Velocity、Tiles、iText 和 POI。Spring MVC 框架并不关心使用的视图技术，也不会强迫开发者只使用 JSP 技术，但本书使用的视图是 JSP。

Spring MVC 框架主要由 DispatcherServlet、处理器映射、控制器、视图解析器、视图组成，其工作原理如图 1.2 所示。

从图 1.2 可总结出 Spring MVC 的工作流程如下：

（1）客户端请求提交到 DispatcherServlet；

（2）由 DispatcherServlet 控制器寻找一个或多个 HandlerMapping（处理器映射），找到处理请求的 Controller；

（3）DispatcherServlet 将请求提交到 Controller；

（4）Controller 调用业务逻辑处理后，返回 ModelAndView；

（5）DispatcherServlet 寻找一个或多个 ViewResolver（视图解析器），找到 ModelAndView 指定的视图；

（6）视图负责将结果显示到客户端。

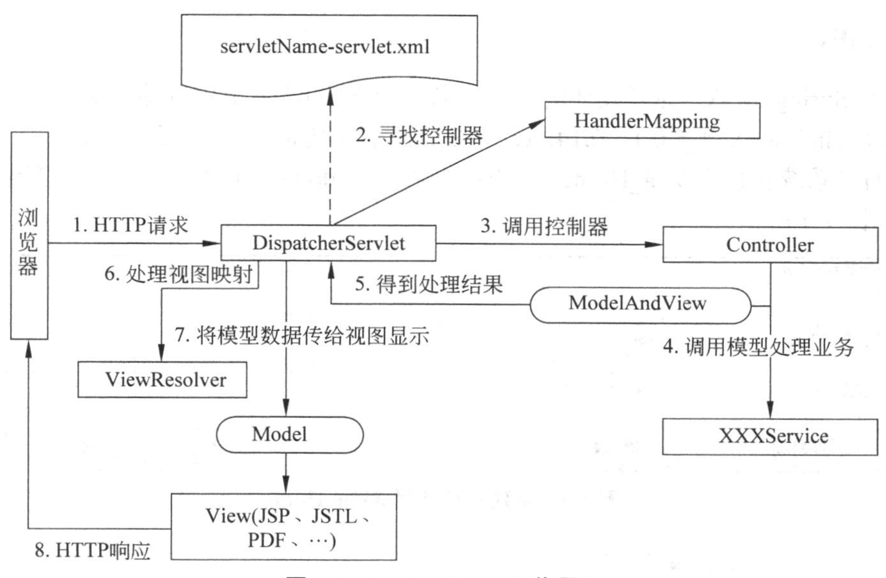

图 1.2　Spring MVC 工作原理

1.1.3　Spring MVC 接口

图 1.2 中包含 4 个 Spring MVC 接口：DispatcherServlet、HandlerMapping、Controller 和 ViewResolver。

Spring MVC 所有的请求都经过 DispatcherServlet 来统一分发。DispatcherServlet 将请求分发给 Controller 之前，需要借助于 Spring MVC 提供的 HandlerMapping 定位到具体的 Controller。

HandlerMapping 接口负责完成客户请求到 Controller 的映射。

Controller 接口将处理用户请求，这和 Java Servlet 扮演的角色是一致的。一旦 Controller 处理完用户请求，则返回 ModelAndView 对象给 DispatcherServlet 前端控制器，ModelAndView 中包含了模型（Model）和视图（View）。从宏观角度考虑，DispatcherServlet 是整个 Web 应用的控制器；从微观角度考虑，Controller 是单个 HTTP 请求处理过程中的控制器，而 ModelAndView 是 HTTP 请求过程中返回的模型（Model）和视图（View）。

ViewResolver 接口在 Web 应用中负责查找 View 对象，从而将相应结果渲染给客户。

1.2　Spring MVC 的开发环境

在第 1.1 节的 MVC 模式中，包含了一个充当调度员的 Servlet；而 Spring MVC 是一个包含 DispatcherServlet 的 MVC 框架，开发者无须编写自己的 Servlet。在使用 Spring MVC 框架进行 Web 开发前，需要构建其开发环境，首先安装 JDK 和 Web 服务器。

1. JDK

构建 Spring MVC 的开发环境，需要首先安装并配置 JDK（本书采用的 JDK 是 jdk-8u111-windows-x64_8.0.1110.14.exe）。按照提示安装完 JDK 之后，需要配置“环境变量”中的“系统变量”Java_Home 和 Path。在 Windows 10 系统下，系统变量示例如图 1.3 和图 1.4 所示。

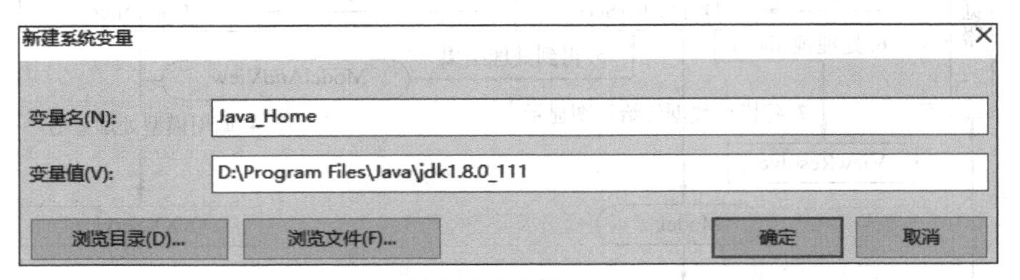

图 1.3 新建系统变量 Java_Home

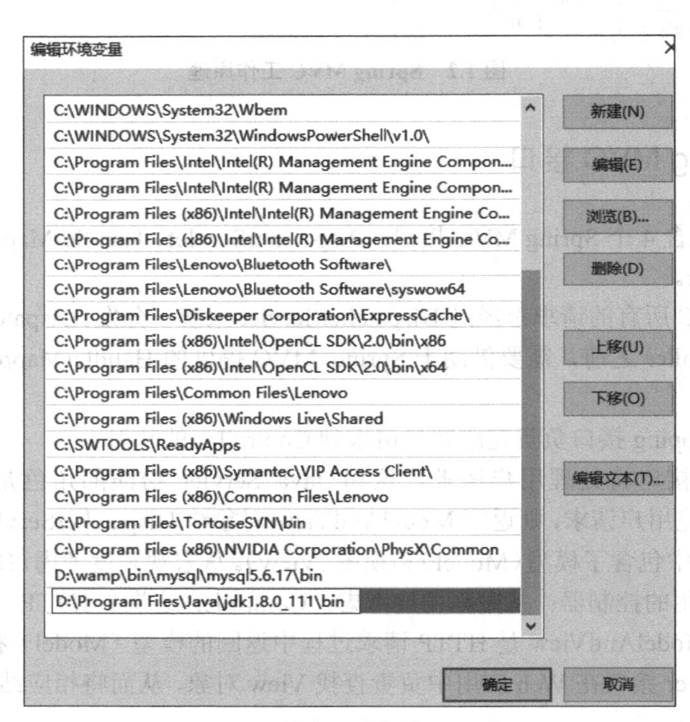

图 1.4 新建环境变量 Path 值

2. Web 服务器

目前，比较常用的 Web 服务器包括 Tomcat、JRun、Resin、WebSphere、WebLogic 等，本书采用的是 Tomcat 8.5。

登录 Apache 软件基金会的官方网站 http://jakarta.Apache.org/tomcat，下载 Tomcat 8.5 的免安装版（apache-tomcat-8.5.11.zip）。登录网站后，首先在 Download 里选择 Tomcat 8，

然后在 Binary Distributions 的 Core 中选择 zip 即可。

安装 Tomcat 之前需要事先安装 JDK 并配置系统环境变量 Java_Home。将下载的 apache-tomcat-8.5.11.zip 解压到磁盘的某个分区中，比如解压到 D:\，解压缩后将出现如图 1.5 所示的目录结构。

D:\apache-tomcat-8.5.11

名称	修改日期	类型	大小
bin	2017/1/10 21:04	文件夹	
conf	2017/1/24 6:02	文件夹	
lib	2017/1/10 21:04	文件夹	
logs	2017/1/24 6:02	文件夹	
temp	2017/1/10 21:04	文件夹	
webapps	2017/1/10 21:04	文件夹	
work	2017/1/24 6:02	文件夹	
LICENSE	2017/1/10 21:04	文件	57 KB
NOTICE	2017/1/10 21:04	文件	2 KB
RELEASE-NOTES	2017/1/10 21:04	文件	8 KB
RUNNING.txt	2017/1/10 21:04	文本文档	17 KB

图 1.5　Tomcat 目录结构

执行 Tomcat 根目录中 bin 文件夹中的 startup.bat 来启动 Tomcat 服务器。执行 startup.bat 启动 Tomcat 服务器会占用一个 MS-DOS 窗口，出现如图 1.6 所示的界面，关闭该 MS-DOS 窗口将关闭 Tomcat 服务器。

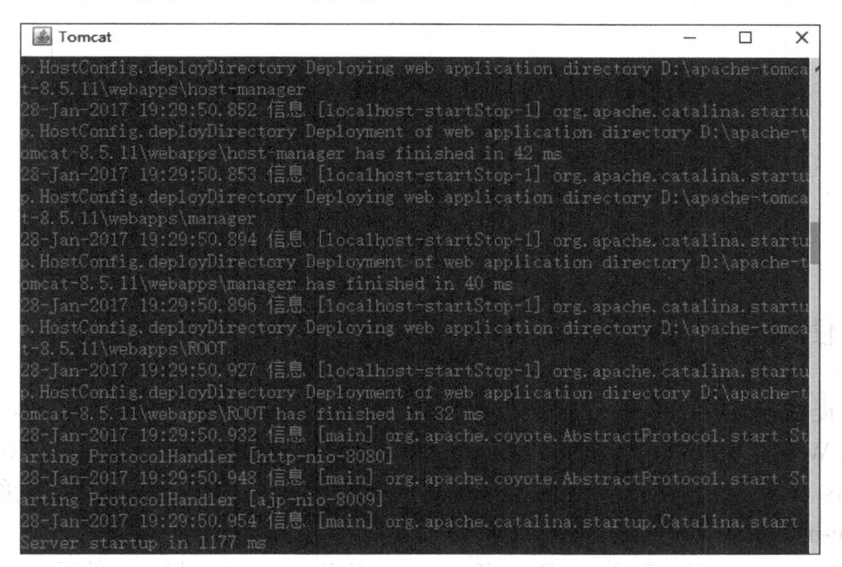

图 1.6　执行 startup.bat 启动 Tomcat 服务器

Tomcat 服务器启动后，在浏览器的地址栏中输入“http://localhost:8080”，将出现如图 1.7 所示的 Tomcat 测试页面。

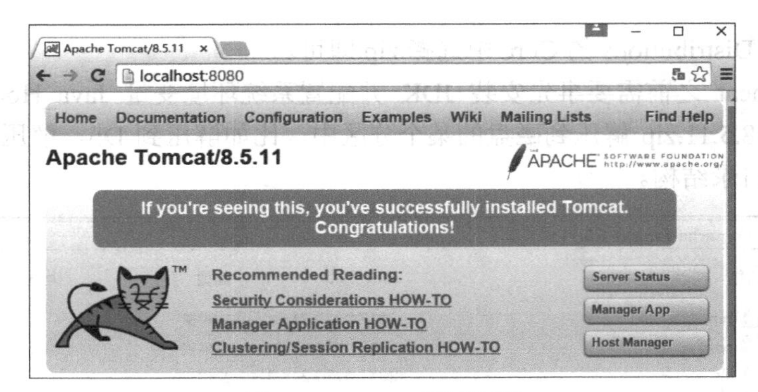

图 1.7 Tomcat 测试页面

1.2.1 Spring 的下载与安装

Spring 官方网站升级后，建议都是通过 Maven 和 Gradle 下载，而不使用 Maven 和 Gradle 的开发者下载 Spring 就非常麻烦。本书给出一个 Spring Framework jar 官方直接下载路径：http://repo.springsource.org/libs-release-local/org/springframework/spring/。本书采用的是 spring-framework-4.3.5.RELEASE-dist.zip。将下载的 ZIP 文件解压缩，解压缩后的目录结构如图 1.8 所示。

D:\spring-framework-4.3.5.RELEASE

名称	修改日期	类型	大小
docs	2016/12/21 11:09	文件夹	
libs	2016/12/21 11:09	文件夹	
schema	2016/12/21 11:09	文件夹	
license.txt	2016/12/21 10:49	文本文档	15 KB
notice.txt	2016/12/21 10:49	文本文档	1 KB
readme.txt	2016/12/21 10:49	文本文档	1 KB

图 1.8 spring-framework-4.3.5 的目录结构

1.2.2 使用 Eclipse 开发 Spring MVC 应用

为了提高开发效率，通常还需要安装 IDE（集成开发环境）工具。Eclipse 是一个可用于开发 Web 应用的 IDE 工具。登录 http://www.eclipse.org，选择 Eclipse IDE for Java EE Developers，根据操作系统的位数，下载相应的 Eclipse。本书采用的是 eclipse-jee-neon-2-win32-x86_64.zip。

使用 Eclipse 之前，需要对 JDK、Tomcat 和 Eclipse 进行一些必要的配置。因此，在安装 Eclipse 之前，应该事先安装 JDK 和 Tomcat。

1. 安装 Eclipse

Eclipse 下载完成后，解压到自己设置的路径下，即可完成安装。Eclipse 安装后，双击 Eclipse 安装目录下的 eclipse.exe 文件，启动 Eclipse。

2．Tomcat 在 Eclipse 中的配置

（1）配置 Tomcat。启动 Eclipse，选择 Window→Preferences 菜单项，在弹出的对话框中选择 Server→Runtime Environments 命令，如图 1.9 所示。

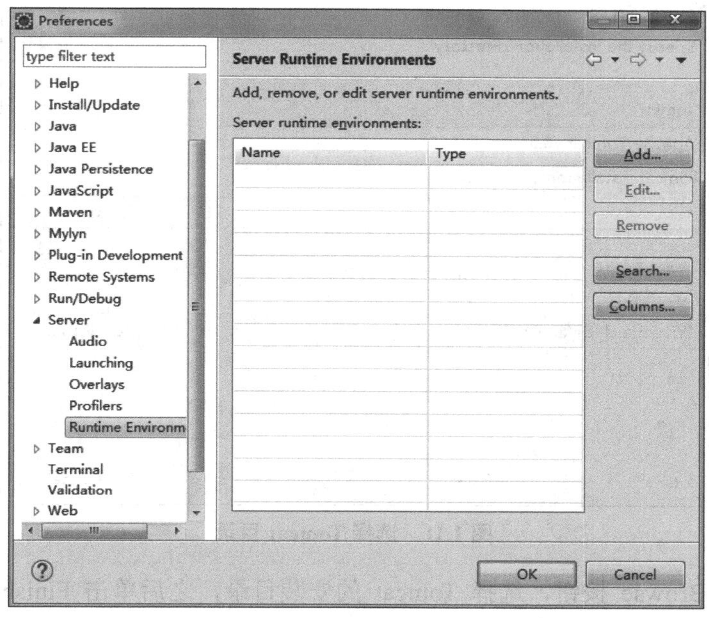

图 1.9 Tomcat 配置界面

（2）单击 Add 按钮后，弹出如图 1.10 所示的 New Server Runtime Environment 界面，在此可以配置各种版本的 Web 服务器。

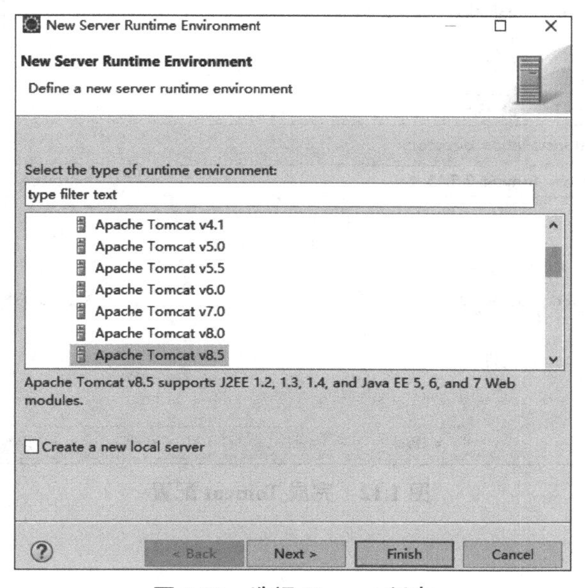

图 1.10 选择 Tomcat 版本

（3）选择 Apache Tomcat v8.5 服务器版本，单击 Next 按钮，进入如图 1.11 所示的界面。

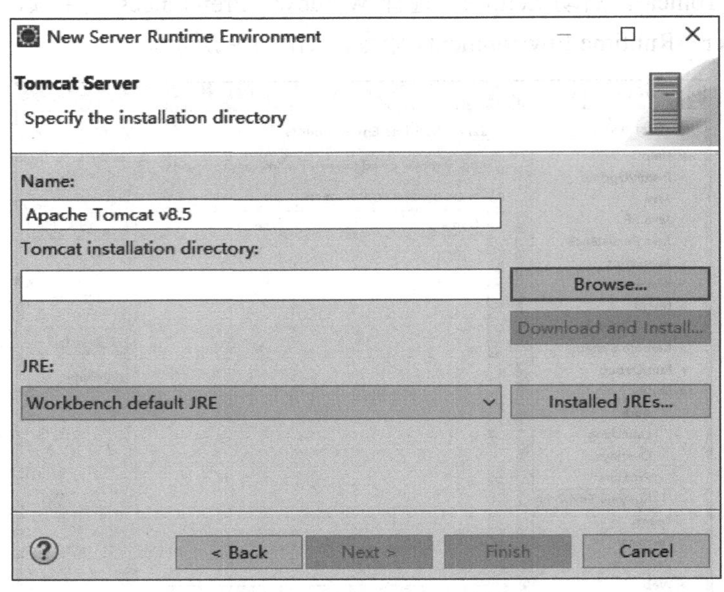

图 1.11　选择 Tomcat 目录

（4）单击 Browse 按钮，选择 Tomcat 的安装目录，之后单击 Finish 按钮即可完成 Tomcat 的配置，如图 1.12 所示。

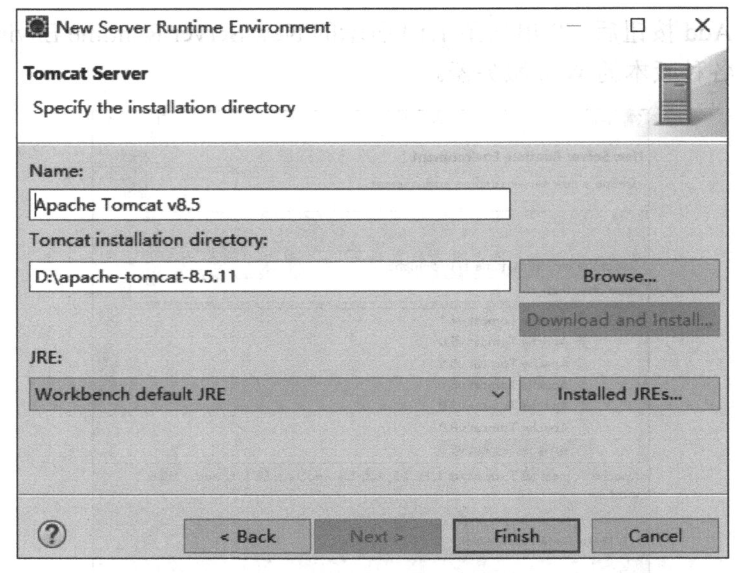

图 1.12　完成 Tomcat 配置

3. 使用 Eclipse 开发 Spring MVC 应用

使用 Eclipse 开发一个 Spring MVC 应用需要如下几个步骤：

① 创建 Web 应用；

② 为 Web 应用添加 Spring MVC 相关类库；

③ 在 web.xml 文件中部署 DispatcherServlet；

④ 创建 Spring MVC 配置文件。

（1）创建 Web 应用

① 启动 Eclipse，进入 Eclipse 开发界面。

② 选择主菜单中的 File→New→Project 菜单项，打开 New Project 对话框，在该对话框中选择 Web 节点下的 Dynamic Web Project 子节点，如图 1.13 所示。

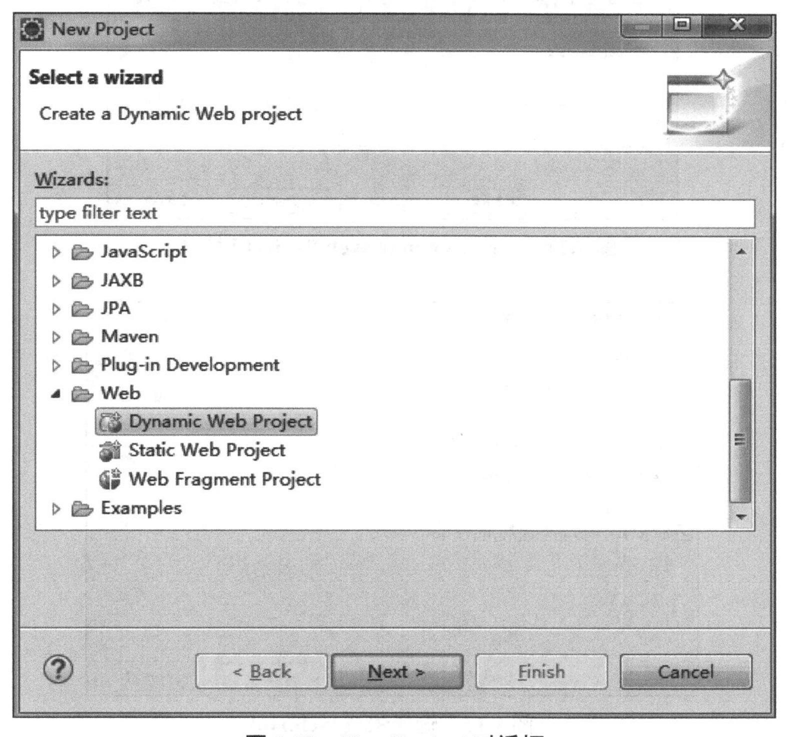

图 1.13 New Project 对话框

③ 单击 Next 按钮，打开 New Dynamic Web Project 对话框，在该对话框的 Project name 文本框中输入应用名称，这里为“firstSpringMVC”。选择 Target runtime 区域中的服务器，如图 1.14 所示。

④ 单击两次 Next 按钮后，选中 Generate web.xml deployment descriptor 选项，如图 1.15 所示。

⑤ 单击 Finish 按钮，完成应用 firstSpringMVC 的创建。此时在 Eclipse 平台左侧，将显示项目 firstSpringMVC，依次展开各节点，可显示如图 1.16 所示的目录结构。

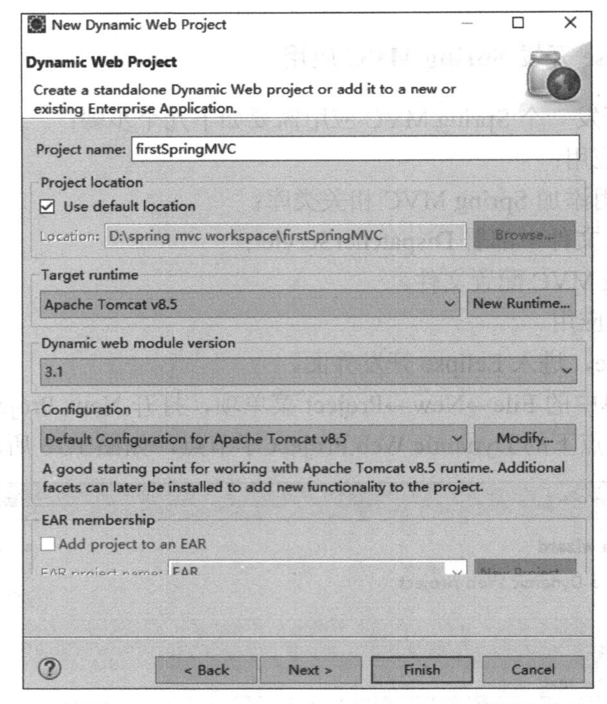

图 1.14　New Dynamic Web Project 对话框

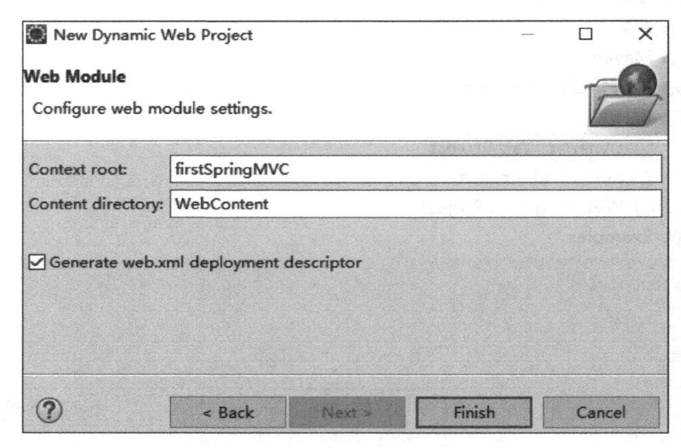

图 1.15　选择生成 web.xml

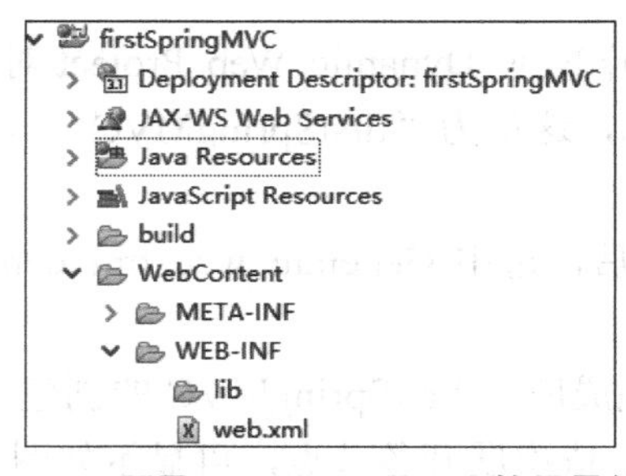

图 1.16　项目 firstSpringMVC 的目录结构

（2）为 Web 应用添加 Spring MVC 相关类库

将 spring-framework-4.3.5.RELEASE-dist.zip 解压缩后的 libs 目录下的 JAR 类库（XXX-4.3.5.RELEASE.jar）复制到 Web 应用的 WEB-INF/lib 目录下。另外，还需要注意的是，Spring MVC 依赖于 Apache Commons Logging 组件，没有该组件，Spring MVC 程序将无法运行。可以从以下网址下载该组件：http://commons.apache.org/proper/commons-logging/download_logging.cgi。本书下载的是 commons-logging-1.2-bin.zip。解压缩后，将 commons-logging-1.2.jar 复制到 Web 应用的 WEB-INF/lib 目录下。

（3）在 web.xml 文件中部署 DispatcherServlet

在开发 Spring MVC 应用时，还需要在 web.xml 中部署 DispatcherServlet，代码如下所示：

```
<?xml version="1.0" encoding="UTF-8"?>
<web-app
xmlns:xsi="http://www.w3.org/2001/XMLSchema-instance"
xmlns="http://xmlns.jcp.org/xml/ns/javaee"
xsi:schemaLocation="http://xmlns.jcp.org/xml/ns/javaee
http://xmlns.jcp.org/xml/ns/javaee/web-app_3_1.xsd"
id="WebApp_ID" version="3.1">
  <display-name>firstSpringMVC</display-name>
  <welcome-file-list>
    <welcome-file>index.html</welcome-file>
    <welcome-file>index.htm</welcome-file>
    <welcome-file>index.jsp</welcome-file>
    <welcome-file>default.html</welcome-file>
    <welcome-file>default.htm</welcome-file>
    <welcome-file>default.jsp</welcome-file>
  </welcome-file-list>
<!--部署 DispatcherServlet-->
<servlet>
    <servlet-name>springmvc</servlet-name>
    <servlet-class>org.springframework.web.servlet.DispatcherServlet
    </servlet-class>
    <load-on-startup>1</load-on-startup>
</servlet>
<servlet-mapping>
    <servlet-name>springmvc</servlet-name>
    <!—处理所有 URL-->
    <url-pattern>/</url-pattern>
</servlet-mapping>
</web-app>
```

上述 DispatcherServlet 的 servlet 对象 springmvc 初始化时，将在应用程序的 WEB-INF 目录下查找一个配置文件，该配置文件的命名规则是“servletName-servlet.xml”，如 springmvc-servlet.xml。

另外，也可以将 Spring MVC 配置文件存放在应用程序目录中的任何地方，不过此时需要使用 servlet 的 init-param 元素加载配置文件。示例代码如下：

```
<!--部署 DispatcherServlet-->
<servlet>
    <servlet-name>springmvc</servlet-name>
    <servlet-class>org.springframework.web.servlet.DispatcherServlet
    </servlet-class>
    <init-param>
         <param-name>contextConfigLocation</param-name>
         <param-value>/WEN-INF/spring-config/springmvc-servlet.xml
          </param-value>
    </init-param>
    <load-on-startup>1</load-on-startup>
</servlet>
<servlet-mapping>
    <servlet-name>springmvc</servlet-name>
    <url-pattern>/</url-pattern>
</servlet-mapping>
```

（4）创建 Spring MVC 配置文件

针对（3），在应用的 WEB-INF 目录下创建配置文件 springmvc-servlet.xml。在 Spring MVC 配置文件中，声明与 Spring MVC 应用相关的配置信息。

经过上面 4 个步骤，已经可以在 firstSpringMVC 应用中使用 Spring MVC 的基本功能了。

1.3 第一个 Spring MVC 应用

本节使用 IDE 开发工具 Eclipse 创建 Web 应用 HelloWorld。在 Eclipse 中如何创建应用、如何添加相关类库到 HelloWorld 的 lib 目录下、如何在 web.xml 文件中部署 DispatcherServlet 以及如何创建 Spring MVC 配置文件，请参考第 1.2.2 节的内容。本节涉及的 Controller 以及相关配置将在后续章节中详细介绍。Spring MVC 应用 HelloWorld 的项目结构如图 1.17 所示。

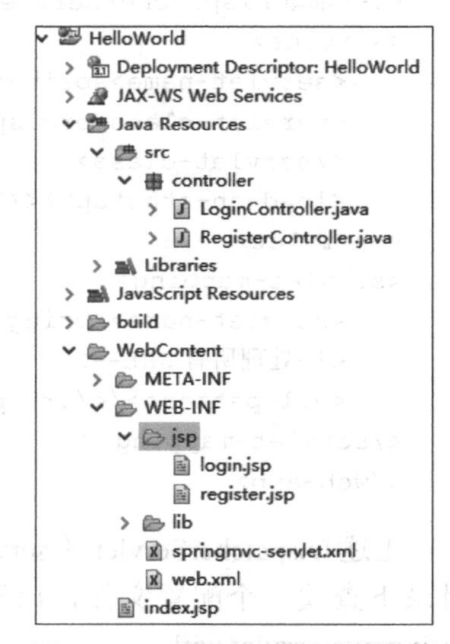

图 1.17 HelloWorld 的项目结构

1.3.1 应用首页

在 HelloWorld 应用的 WebContent 目录下，有个应用首页 index.jsp。index.jsp 的代码如下：

```
<%@ page language="java" contentType="text/html; charset=UTF-8"
  pageEncoding = "UTF-8"%>
<%
String path = request.getContextPath();
String basePath = request.getScheme()+"://"+request.getServerName()+":
"+request.getServerPort()+path+"/";
%>
<!DOCTYPE html PUBLIC "-//W3C//DTD HTML 4.01 Transitional//EN"
"http://www.w3.org/TR/html4/loose.dtd">
<html>
  <head>
     <base href="<%=basePath%>">
    <meta http-equiv="Content-Type" content="text/html; charset=UTF-8">
    <title>My JSP 'index.jsp' starting page</title>
  </head>
  <body>
    没注册的用户，请<a href="register">注册</a>! <br>
    已注册的用户，去<a href="login">登录</a>!
  </body>
</html>
```

1.3.2 实现 Controller

在 HelloWorld 应用中有 RegisterController 和 LoginController 两个传统风格的控制器，分别处理“注册”和“登录”超链接请求。

RegisterController 的具体代码如下：

```
package controller;
import javax.servlet.http.HttpServletRequest;
import javax.servlet.http.HttpServletResponse;
import org.springframework.web.servlet.ModelAndView;
import org.springframework.web.servlet.mvc.Controller;
public class RegisterController implements Controller{
    @Override
    public ModelAndView handleRequest(HttpServletRequest arg0,
HttpServletResponse arg1) throws Exception {
        return new ModelAndView("/WEB-INF/jsp/register.jsp");
    }
}
```

LoginController 的具体代码如下：

```
package controller;
import javax.servlet.http.HttpServletRequest;
import javax.servlet.http.HttpServletResponse;
import org.springframework.web.servlet.ModelAndView;
```

```
import org.springframework.web.servlet.mvc.Controller;
public class LoginController implements Controller{
    @Override
    public ModelAndView handleRequest(HttpServletRequest arg0,
HttpServletResponse arg1) throws Exception {
        return new ModelAndView("/WEB-INF/jsp/login.jsp");
    }
}
```

1.3.3 配置 Controller

传统风格的控制器在定义后，还需要在 Spring MVC 配置文件中部署它们（学习基于注解的控制器后，就不再需要部署控制器了），具体代码如下：

```
<?xml version="1.0" encoding="UTF-8"?>
<beans xmlns="http://www.springframework.org/schema/beans"
    xmlns:xsi="http://www.w3.org/2001/XMLSchema-instance"
    xsi:schemaLocation="
        http://www.springframework.org/schema/beans
        http://www.springframework.org/schema/beans/spring-beans.xsd">
  <!--LoginController 控制器类，映射到“/login”  -->
  <bean name="/login" class="controller.LoginController"/>
  <!--RegisterController 控制器类，映射到“/register”  -->
  <bean name="/register" class="controller.RegisterController"/>
</beans>
```

1.3.4 应用的其他页面

RegisterController 控制器处理成功后，跳转到/WEB-INF/jsp/register.jsp 视图；LoginController 控制器处理成功后，跳转到/WEB-INF/jsp/login.jsp 视图。因此，应用的/WEB-INF/jsp 目录下应有 register.jsp 和 login.jsp 页面。此两个 JSP 页面的代码这里略过。

1.3.5 发布并运行 Spring MVC 应用

在 Eclipse 中第一次运行 Spring MVC 应用时，需要将应用发布到 Tomcat。例如，运行 HelloWorld 应用时，可以选中应用名称 HelloWorld 右击，选择 Run As→Run on Server，打开如图 1.18 所示的对话框，在对话框中单击 Finish 按钮即完成发布并运行。

此后，通过地址 http://localhost:8080/HelloWorld 将首先访问 index.jsp 页面，如图 1.19 所示。

在如图 1.19 所示的页面中，用户单击“注册”超链接时，根据 springmvc-servlet.xml 文件中的映射，请求将被转发给 RegisterController 控制器处理，处理后跳转到/WEB-INF

/jsp/register.jsp 视图。同理，单击“登录”超链接时，控制器处理后转到/WEB-INF/jsp/login.jsp 视图。

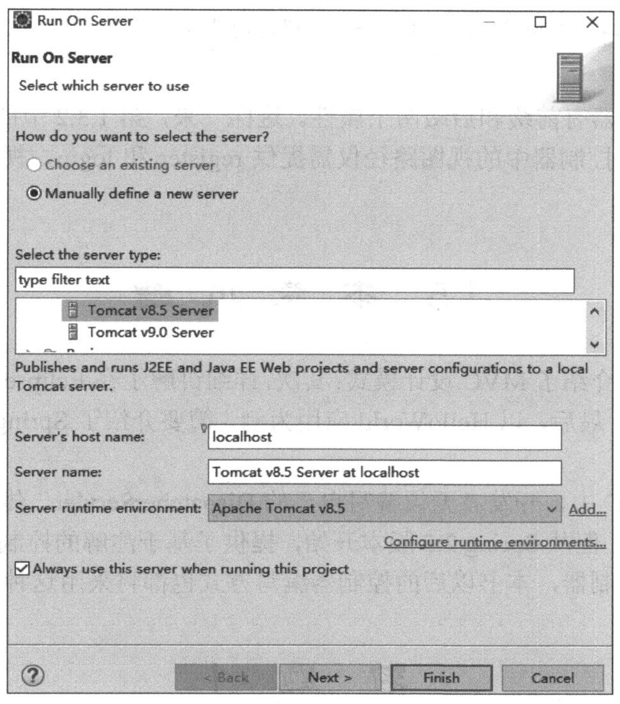

图 1.18 在 Eclipse 中发布并运行 Spring MVC 应用

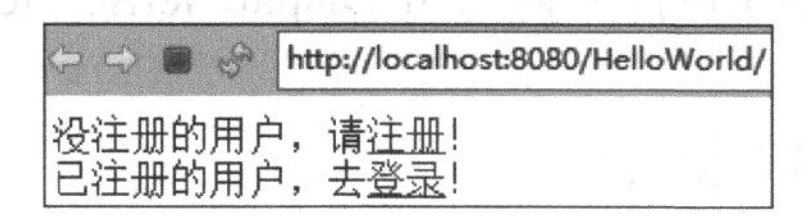

图 1.19 index.jsp 页面

1.3.6 实践环节

（1）下载 Spring 最新版本。

（2）在 Eclipse 中为 Web 应用搭建 Spring MVC 的开发环境。

1.4 视图解析器

可以在配置文件中定义 Spring MVC 的一个视图解析器（ViewResolver），示例代码如下：

```
<bean class="org.springframework.web.servlet.view.
    InternalResourceViewResolver" id="internalResourceViewResolver">
```

```
    <!-- 前缀 -->
    <property name="prefix" value="/WEB-INF/jsp/" />
    <!-- 后缀 -->
    <property name="suffix" value=".jsp" />
</bean>
```

上述视图解析器有前缀和后缀两个属性。这样一来，第 1.3.2 节的 RegisterController 和 LoginController 控制器中的视图路径仅需提供 register 和 login，视图解析器将会自动添加前缀和后缀。

1.5 本章小结

本章首先简单介绍了 MVC 设计模式；其次，详细讲解了在 Eclipse 中如何构建 Spring MVC 的开发环境；最后，以 HelloWorld 应用为例，简要介绍了 Spring MVC 框架的基本流程。

在 Spring MVC 中，开发者无须编写自己的 DispatcherServlet。传统的控制器需要实现 Controller 接口。但从 Spring 2.5 版本开始，提供了基于注解的控制器。第 2 章会详细介绍基于注解的控制器，本书以后的控制器编写方式也都将采用这种注解的方式。

习题 1

1. 在开发 Spring MVC 应用时，如何部署 DispatcherServlet？又如何创建 Spring MVC 的配置文件？

2. 简述 Spring MVC 的工作流程。

第2章

Controller

学习目的与要求

本章重点讲解基于注解的控制器、Controller 接收请求参数的方式，以及如何编写请求处理方法。通过本章的学习，读者应掌握基于注解的控制器的编写方法，掌握在 Controller 中如何接收请求参数以及编写请求处理方法。

本章主要内容

- 基于注解的控制器
- 编写请求处理方法
- Controller 接收请求参数的方式
- 重定向和转发
- 应用@Autowired 和@Service 进行依赖注入
- @ModelAttribute

在使用 Spring MVC 进行 Web 应用开发时，Controller 是 Web 应用的核心。Controller 实现类包含了对用户请求的处理逻辑，是用户请求和业务逻辑之间的“桥梁”，是 Spring MVC 框架的核心部分，负责具体的业务逻辑处理。

2.1 基于注解的控制器

在第 1.3 节“第一个 Spring MVC 应用”中创建了两个传统风格的控制器，它们是实现了 Controller 接口的类。传统风格的控制器不仅需要在配置文件中部署映射，而且只能编写一个处理方法，不够灵活。使用基于注解的控制器，具有如下两个优点：

（1）在基于注解的控制器类中，可以编写多个处理方法，进而可以处理多个请求（动作）。这就允许将相关的操作编写在同一个控制器类中，从而减少控制器类的数量，方便以后的维护。

（2）基于注解的控制器不需要在配置文件中部署映射，仅需要使用 RequestMapping 注释类型注解一个方法进行请求处理。

在 Spring MVC 中，最重要的两个注释类型是 Controller 和 RequestMapping 注释类

型，本章重点介绍它们。本章将创建一个 Spring MVC 应用 ch2，ch2 的项目结构如图 2.1 所示。

图 2.1 ch2 的项目结构

2.1.1 Controller 注解类型

在 Spring MVC 中，使用 org.springframework.stereotype.Controller 注解类型声明某类的实例是一个控制器。Controller 注解示例代码如下：

```
package controller;
import org.springframework.stereotype.Controller;
//“@Controller”表示 IndexController 的实例是一个控制器
@Controller
public class IndexController {
//处理请求的方法
}
```

Spring MVC 使用扫描机制找到应用中所有基于注解的控制器类。所以，为了让控制器类能被 Spring MVC 框架扫描到，需要在配置文件中声明 spring-context，并使用 <context:component-scan/>元素指定控制器类的基本包（请确保所有控制器类都在基本包及其子包下）。配置文件具体示例代码如下：

```
<?xml version="1.0" encoding="UTF-8"?>
<beans xmlns="http://www.springframework.org/schema/beans"
    xmlns:xsi="http://www.w3.org/2001/XMLSchema-instance"
    xmlns:p="http://www.springframework.org/schema/p"
    xmlns:context="http://www.springframework.org/schema/context"
    xsi:schemaLocation="
```

```
    http://www.springframework.org/schema/beans
    http://www.springframework.org/schema/beans/spring-beans.xsd
       http://www.springframework.org/schema/context
       http://www.springframework.org/schema/context/
       spring-context.xsd">
   <!-- 使用扫描机制，扫描控制器类，控制器类都在 controller 包及其子包下 -->
   <context:component-scan base-package="controller"/>
    <!-- … -->
</beans>
```

2.1.2 RequestMapping 注解类型

在基于注解的控制器类中，可以为每个请求编写对应的处理方法。如何将请求与处理方法一一对应呢？需要使用 org.springframework.web.bind.annotation.RequestMapping 注解类型。

1. 方法级别注解

方法级别注解示例代码如下：

```
package controller;
import org.springframework.stereotype.Controller;
import org.springframework.web.bind.annotation.RequestMapping;
@Controller
public class IndexController {
     @RequestMapping(value = "/index/login")
     public String login() {
          return "login";//login 代表逻辑视图名称
     }
     @RequestMapping(value = "/index/register")
     public String register() {
          return "register";
     }
}
```

上述示例中有两个 RequestMapping 注解语句，它们都作用在处理方法上。注解的 value 属性将请求 URI 映射到方法。value 属性是 RequestMapping 注解的默认属性，如果只有一个 value 属性，则可省略该属性。可以使用如下 URL 访问 login 方法（请求处理方法）。

http://localhost:8080/ch2/index/login

2. 类级别注解

类级别注解示例代码如下：

```
package controller;
import org.springframework.stereotype.Controller;
import org.springframework.web.bind.annotation.RequestMapping;
@Controller
@RequestMapping("/index")
public class IndexController {
      @RequestMapping("/login")
      public String login() {
            return "login";
      }
      @RequestMapping("/register")
      public String register() {
           return "register";
      }
}
```

在类级别注解的情况下，控制器类中的所有方法都将映射为类级别的请求。可以使用如下 URL 访问 login 方法。

http://localhost:8080/ch2/index/login

为了方便程序维护，建议开发者采用类级别注解，将相关处理放在同一个控制器类中，例如，对商品的增、删、改、查等处理方法都可以放在 GoodsOperate 控制类中。

2.1.3 编写请求处理方法

在控制器类中的每个请求处理方法可以有多个不同类型的参数，以及一个多种类型的返回结果。

1. 请求处理方法中常出现的参数类型

如果需要在请求处理方法中使用 Servlet API 类型，那么可以将这些类型作为请求处理方法的参数类型。Servlet API 参数类型示例代码如下：

```
package controller;
import javax.servlet.http.HttpServletRequest;
import javax.servlet.http.HttpSession;
import org.springframework.stereotype.Controller;
import org.springframework.web.bind.annotation.RequestMapping;
@Controller
@RequestMapping("/index")
public class IndexController {
      @RequestMapping("/login")
     public String login(HttpSession session, HttpServletRequest request){
           session.setAttribute("skey", "session 范围的值");
           request.setAttribute("rkey", "request 范围的值");
           return "login";
```

```
    }
}
```

除了 Servlet API 参数类型外，还有输入输出流、表单实体类、注解类型、与 Spring 框架相关的类型等，这些类型在后续章节中使用时再详细介绍。但特别重要的类型是 org.springframework.ui.Model 类型，该类型是一个包含 Map 的 Spring 框架类型。每次调用请求处理方法时，Spring MVC 都将创建 org.springframework.ui.Model 对象。Model 参数类型示例代码如下：

```
package controller;
import org.springframework.stereotype.Controller;
import org.springframework.ui.Model;
import org.springframework.web.bind.annotation.RequestMapping;
@Controller
@RequestMapping("/index")
public class IndexController {
    @RequestMapping("/register")
    public String register(Model model) {
            /*在视图中可以使用 EL 表达式${success}取出 model 中的值,有关 EL 的
相关知识,请参考本书有关内容。*/
        model.addAttribute("success", "注册成功");
        return "register";
    }
}
```

2．请求处理方法常见的返回类型

最常见的返回类型就是代表逻辑视图名称的 String 类型，如前面章节中的请求处理方法。除了 String 类型外，还有 ModelAndView（如第 1 章的传统控制器）、Model、View 以及其他任意的 Java 类型。

2.2 Controller 接收请求参数的常见方式

Controller 接收请求参数的方式有很多种，有的适合 get 请求方式，有的适合 post 请求方式，有的两者都适合。下面分别介绍这些方式，读者可根据实际情况选择合适的接收方式。

2.2.1 通过实体 bean 接收请求参数

通过一个实体 bean 来接收请求参数，适用于 get 和 post 提交请求方式。需要注意的是，bean 的属性名称必须与请求参数名称相同。下面通过具体应用 ch2 来讲解如何通过实体 bean 接收请求参数。

应用 ch2 的首页 index.jsp 的代码如下：

```
<%@ page language="java" contentType="text/html; charset=UTF-8"
pageEncoding="UTF-8"%>
<%
String path = request.getContextPath();
String basePath = request.getScheme()+"://"+request.getServerName()+":"+
request.getServerPort()+path+"/";
%>
<!DOCTYPE html PUBLIC "-//W3C//DTD HTML 4.01 Transitional//EN"
"http://www.w3.org/TR/html4/loose.dtd">
<html>
  <head>
     <base href="<%=basePath%>">
    <meta http-equiv="Content-Type" content="text/html; charset=UTF-8">
    <title>My JSP 'index.jsp' starting page</title>
  </head>
  <body>
    没注册的用户，请<a href="index/register">注册</a>! <br>
    已注册的用户，去<a href="index/login">登录</a>!
  </body>
</html>
```

应用 ch2 的配置文件 springmvc-servlet.xml 的代码如下：

```
<?xml version="1.0" encoding="UTF-8"?>
<beans xmlns="http://www.springframework.org/schema/beans"
    xmlns:xsi="http://www.w3.org/2001/XMLSchema-instance"
    xmlns:p="http://www.springframework.org/schema/p"
    xmlns:context="http://www.springframework.org/schema/context"
    xmlns:mvc="http://www.springframework.org/schema/mvc"
    xsi:schemaLocation="
     http://www.springframework.org/schema/beans
     http://www.springframework.org/schema/beans/spring-beans.xsd
        http://www.springframework.org/schema/context
        http://www.springframework.org/schema/context/spring-context.xsd
        http://www.springframework.org/schema/mvc
        http://www.springframework.org/schema/mvc/spring-mvc.xsd">
    <!-- 使用扫描机制，扫描控制器类 -->
    <context:component-scan base-package="controller"/>
    <mvc:annotation-driven />
    <!-- annotation-driven 用于简化开发的配置，
    注解 DefaultAnnotationHandlerMapping 和 AnnotationMethodHandlerAdapter
-->
    <!-- 使用 resources 过滤掉不需要 dispatcher servlet 的资源。
    使用 resources 时，必须使用 annotation-driven，不然 resources 元素会阻止任意
控制器被调用。
```

```
    如果不使用 resources，则 annotation-driven 可以没有。 -->
    <!-- 允许 css 目录下所有文件可见 -->
    <mvc:resources location="/css/" mapping="/css/**"></mvc:resources>
    <!-- 允许 html 目录下所有文件可见 -->
    <mvc:resources location="/html/" mapping="/html/**"></mvc:resources>
     <!--允许 images 目录下所有文件可见 -->
    <mvc:resources location="/images/" mapping="/images/**">
    </mvc:resources>
     <!-- 配置视图解析器 -->
     <bean class="org.springframework.web.servlet.view.
                                          InternalResourceViewResolver"
           id="internalResourceViewResolver">
       <!-- 前缀 -->
       <property name="prefix" value="/WEB-INF/jsp/" />
       <!-- 后缀 -->
       <property name="suffix" value=".jsp" />
  </bean>
</beans>
```

应用 ch2 的实体 bean 类 UserForm 的代码如下：

```
package domain;
public class UserForm {
    private String uname;//与请求参数名称相同
    private String upass;
    private String reupass;
    public String getUname() {
          return uname;
    }
    public void setUname(String uname) {
          this.uname = uname;
    }
    public String getUpass() {
          return upass;
    }
    public void setUpass(String upass) {
          this.upass = upass;
    }
    public String getReupass() {
          return reupass;
    }
    public void setReupass(String reupass) {
          this.reupass = reupass;
    }
}
```

应用 ch2 的控制器类 IndexController 的代码如下：

```
package controller;
import org.springframework.stereotype.Controller;
import org.springframework.web.bind.annotation.RequestMapping;
@Controller
@RequestMapping("/index")
public class IndexController {
     @RequestMapping("/login")
     public String login() {
          return "login";//跳转到"/WEB-INF/jsp/login.jsp"
     }
     @RequestMapping("/register")
     public String register() {
        return "register";
     }
}
```

应用 ch2 的控制器类 UserController 的代码如下：

```
package controller;
import org.springframework.stereotype.Controller;
import org.springframework.ui.Model;
import org.springframework.web.bind.annotation.RequestMapping;
import domain.UserForm;
@Controller
@RequestMapping("/user")
public class UserController {
    @RequestMapping("/register")
    /**
     * UserForm 对象 user 接收注册页面提交的请求参数
     */
    public String register(UserForm user, Model model) {
         if("zhangsan".equals(user.getUname())
                   && "123456".equals(user.getUpass()))
              return "login";//注册成功，跳转到 login.jsp
         else{
              //在 register.jsp 页面上可以使用 EL 表达式取出 model 的 uname 值
              model.addAttribute("uname", user.getUname());
              return "register";//返回 register.jsp
         }
    }
}
```

应用 ch2 的 register.jsp 的代码如下：

```
<%@ page language="java" contentType="text/html; charset=UTF-8"
pageEncoding="UTF-8"%>
```

```
<%
String path = request.getContextPath();
String basePath = request.getScheme()+"://"+request.getServerName()+
":"+request.getServerPort()+path+"/";
%>
<!DOCTYPE html PUBLIC "-//W3C//DTD HTML 4.01 Transitional//EN"
"http://www.w3.org/TR/html4/loose.dtd">
<html>
<head>
<base href="<%=basePath%>">
<meta http-equiv="Content-Type" content="text/html; charset=UTF-8">
<style type="text/css">
     .textSize{
          width: 100pt;
          height: 15pt
     }
</style>
<title>注册画面</title>
<script type="text/javascript">
     //注册时检查输入项
     function allIsNull(){
          var name=document.registForm.uname.value;
          var pwd=document.registForm.upass.value;
          var repwd=document.registForm.reupass.value;
          if(name==""){
                 alert("请输入姓名！");
                 document.registForm.uname.focus();
                 return false;
          }
          if(pwd==""){
                 alert("请输入密码！");
                 document.registForm.upass.focus();
                 return false;
          }
          if(repwd==""){
                 alert("请输入确认密码！");
                 document.registForm.reupass.focus();
                 return false;
          }
          if(pwd!=repwd){
                 alert("2 次密码不一致，请重新输入！");
                 document.registForm.upass.value="";
                 document.registForm.reupass.value="";
                 document.registForm.upass.focus();
                 return false;
          }
```

```
            document.registForm.submit();
            return true;
        }
</script>
</head>
<body>
     <form action="user/register" method="post" name="registForm">
          <table  border=1  bgcolor="lightblue" align="center">
               <tr>
                    <td>姓名：</td>
                    <td>
                         <input  class="textSize"  type="text"  name="uname"
value="${uname }"/>
                    </td>
               </tr>
               <tr>
                    <td>密码：</td>
                    <td><input class="textSize" type="password" maxlength=
"20" name="upass"/></td>
               </tr>
               <tr>
                    <td>确认密码：</td>
          <td><input  class="textSize"  type="password"  maxlength="20"
name="reupass"/></td>
               </tr>
               <tr>
          <td colspan="2" align="center"><input type="button" value="注册
" onclick="allIsNull()"/></td>
               </tr>
          </table>
     </form>
</body>
</html>
```

在register.jsp的代码中使用了EL语句“${uname }”取出“model.addAttribute("uname", user.getUname())”中的值。有关EL与JSTL的相关知识，请参考本书的相关内容。

应用ch2的login.jsp的代码如下：

```
<%@ page language="java" contentType="text/html; charset=UTF-8" pageEncoding
="UTF-8"%>
<%
String path = request.getContextPath();
String basePath = request.getScheme()+"://"+request.getServerName()+":"+
request.getServerPort()+path+"/";
%>
<!DOCTYPE html PUBLIC "-//W3C//DTD HTML 4.01 Transitional//EN"
```

```
"http://www.w3.org/TR/html4/loose.dtd">
<html>
  <head>
  <base href="<%=basePath%>">
  <meta http-equiv="Content-Type" content="text/html; charset=UTF-8">
    <title>后台登录</title>
      <style type="text/css">
      table{
          text-align: center;
      }
      .textSize{
          width: 120px;
          height: 25px;
      }
      * {
          margin: 0px;
          padding: 0px;
      }
      body {
         font-family: Arial, Helvetica, sans-serif;
         font-size: 12px;
         margin: 10px 10px auto;
         background-image: url(images/bb.jpg);
      }
      </style>
      <script type="text/javascript">
      //"确定"按钮
      function gogo(){
         document.forms[0].submit();
      }
      //"取消"按钮
      function cancel(){
         document.forms[0].action = "";
      }
      </script>
  </head>
  <body>
     <form action="user/login" method="post">
     <table>
         <tr>
              <td colspan="2"><img src="images/login.gif"></td>
         </tr>
         <tr>
              <td>姓名：</td>
              <td><input type="text" name="uname" class="textSize"></td>
         </tr>
```

```
            <tr>
                <td>密码：</td>
                <td><input type="password" name="upass" class= "textSize">
</td>
            </tr>
            <tr>
                <td colspan="2">
                    <input type="image" src="images/ok.gif" onclick="gogo()">
                    <input type="image" src="images/cancel.gif" onclick=
"cancel()" >
                </td>
            </tr>
        </table>
        </form>
    </body>
</html>
```

2.2.2 通过处理方法的形参接收请求参数

通过处理方法的形参接收请求参数，也就是直接把表单参数写在控制器类相应方法的形参中，即形参名称与请求参数名称完全相同。该接收参数方式适用于 get 和 post 提交请求方式。可以将第 2.2.1 节中控制器类 UserController 的代码修改如下：

```
package controller;
import org.springframework.stereotype.Controller;
import org.springframework.ui.Model;
import org.springframework.web.bind.annotation.RequestMapping;
import domain.UserForm;
@Controller
@RequestMapping("/user")
public class UserController {
    @RequestMapping("/register")
    /**
     * 通过形参接收请求参数，形参名称与请求参数名称完全相同
     */
    public String register(String uname, String upass, Model model) {
        if("zhangsan".equals(uname)
                && "123456".equals(upass))
            return "login";//注册成功，跳转到 login.jsp
        else{
            //在 register.jsp 页面上可以使用 EL 表达式取出 model 的 uname 值
            model.addAttribute("uname", uname);
            return "register";//返回 register.jsp
        }
```

```
    }
}
```

2.2.3 通过 HttpServletRequest 接收请求参数

通过 HttpServletRequest 接收请求参数，适用于 get 和 post 提交请求方式。可以将第 2.2.1 节中控制器类 UserController 的代码修改如下：

```
package controller;
import javax.servlet.http.HttpServletRequest;
import org.springframework.stereotype.Controller;
import org.springframework.ui.Model;
import org.springframework.web.bind.annotation.RequestMapping;
@Controller
@RequestMapping("/user")
public class UserController {
    @RequestMapping("/register")
    /**
     * 通过 HttpServletRequest 接收请求参数
     */
    public String register(HttpServletRequest request, Model model) {
        String uname = request.getParameter("uname");
        String upass = request.getParameter("upass");
        if("zhangsan".equals(uname)
                && "123456".equals(upass))
            return "login";//注册成功，跳转到 login.jsp
        else{
            //在 register.jsp 页面上可以使用 EL 表达式取出 model 的 uname 值
            model.addAttribute("uname", uname);
            return "register";//返回 register.jsp
        }
    }
}
```

2.2.4 通过@PathVariable 接收 URL 中的请求参数

通过@PathVariable 获取 URL 中的参数，控制器类示例代码如下：

```
package controller;
import org.springframework.stereotype.Controller;
import org.springframework.ui.Model;
import org.springframework.web.bind.annotation.PathVariable;
import org.springframework.web.bind.annotation.RequestMapping;
import org.springframework.web.bind.annotation.RequestMethod;
@Controller
```

```
@RequestMapping("/user")
public class UserController {
    @RequestMapping(value="/register/{uname}/{upass}", method=
RequestMethod.GET)
    //必须加 method 属性
    /**
     * 通过@PathVariable 获取 URL 中的参数
     */
    public String register(@PathVariable String uname,@PathVariable
String upass, Model model) {
        if("zhangsan".equals(uname)
                && "123456".equals(upass))
            return "login";//注册成功，跳转到 login.jsp
        else{
            //在 register.jsp 页面上可以使用 EL 表达式取出 model 的 uname 值
            model.addAttribute("uname", uname);
            return "register";//返回 register.jsp
        }
    }
}
```

访问 http://localhost:8080/ch2/user/register/zhangsan/123456 路径时，上述代码自动将 URL 中的模板变量{uname}和{upass}绑定到通过@PathVariable 注解的同名参数上，即 uname=zhangsan、upass=123456。

2.2.5 通过@RequestParam 接收请求参数

通过@RequestParam 接收请求参数，适用于 get 和 post 提交请求方式。可以将第 2.2.1 节中控制器类 UserController 的代码修改如下：

```
package controller;
import org.springframework.stereotype.Controller;
import org.springframework.ui.Model;
import org.springframework.web.bind.annotation.RequestMapping;
import org.springframework.web.bind.annotation.RequestParam;
@Controller
@RequestMapping("/user")
public class UserController {
    @RequestMapping("/register")
    /**
     * 通过@RequestParam 接收请求参数
     */
    public String register(@RequestParam String uname, @RequestParam
String upass, Model model) {
        if("zhangsan".equals(uname)
```

```
                && "123456".equals(upass))
            return "login";//注册成功，跳转到 login.jsp
        else{
            //在 register.jsp 页面上可以使用 EL 表达式取出 model 的 uname 值
            model.addAttribute("uname", uname);
            return "register";//返回 register.jsp
        }
    }
}
```

通过@RequestParam 接收请求参数与第 2.2.2 节“通过处理方法的形参接收请求参数”的区别是：当请求参数名与接收参数名不一致时，通过处理方法的形参接收请求参数不会报 404 错误，而通过@RequestParam 接收请求参数会报 404 错误。

2.2.6 通过@ModelAttribute 接收请求参数

@ModelAttribute 注解放在处理方法的形参上时，用于将多个请求参数封装到一个实体对象，从而简化数据绑定流程，而且自动暴露为模型数据，于视图页面展示时使用。而第 2.2.1 节中只是将多个请求参数封装到一个实体对象，并不能暴露为模型数据（需要使用 model.addAttribute 语句才能暴露为模型数据。关于数据绑定与模型数据展示的内容，可参考第 4 章）。

通过@ModelAttribute 注解接收请求参数，适用于 get 和 post 提交请求方式。可以将第 2.2.1 节中控制器类 UserController 的代码修改如下：

```
package controller;
import org.springframework.stereotype.Controller;
import org.springframework.web.bind.annotation.ModelAttribute;
import org.springframework.web.bind.annotation.RequestMapping;
import domain.UserForm;
@Controller
@RequestMapping("/user")
public class UserController {
    /**
     *处理注册
     */
    @RequestMapping("/register")
    public String register(@ModelAttribute("user") UserForm user) {
        if("zhangsan".equals(user.getUname())
                && "123456".equals(user.getUpass())){
            return "login";//注册成功，跳转到 login.jsp
        }else{
            //使用@ModelAttribute("user")与 model.addAttribute("user",
user)功能相同
        //在register.jsp页面上可以使用EL表达式${user.uname}取出ModelAttribute
```

```
的 uname 值
            return "register";//返回 register.jsp
        }
    }
}
```

2.2.7 实践环节

在第 2.2.1 节的控制器类 UserController 中，添加一个处理登录请求（单击 login.jsp 页面中的“确定”按钮）的方法：

```
@RequestMapping("/login")
public String login(UserForm user, HttpSession session, Model model) {
    …
}
```

在该方法中，判断用户是否登录成功。如果用户名为 zhangsan，密码为 123456，则登录成功，否则失败。登录成功则跳转到 main.jsp 并将用户信息存储在 session 对象中。登录失败则返回登录页面，并将“用户名或密码错误”存储在 model 对象中。

在 main.jsp 页面中使用 EL 表达式取出 session 对象中的用户信息，在 login.jsp 页面中使用 EL 表达式取出 model 对象中的错误消息。

2.3 重定向与转发

重定向是将用户从当前处理请求定向到另一个视图（如 JSP）或处理请求，以前的请求（request）中存放的信息全部失效，并进入一个新的 request 作用域；转发是将用户对当前处理的请求转发给另一个视图或处理请求，以前的 request 中存放的信息不会失效。

转发是服务器行为，重定向是客户端行为。具体工作流程如下。

转发过程：客户浏览器发送 HTTP 请求，Web 服务器接收此请求，调用内部的一个方法在容器内部完成请求处理和转发动作，将目标资源发送给客户。在这里，转发的路径必须是同一个 Web 容器下的 URL，不能转向到其他的 Web 路径上去，中间传递的是自己的容器内的 request。在客户浏览器的地址栏中显示的仍然是其第一次访问的路径，也就是说，客户是感觉不到服务器做了转发的。转发行为是浏览器只做了一次访问请求。

重定向过程：客户浏览器发送 HTTP 请求，Web 服务器接收后发送 302 状态码响应，并将对应的新 location 发给客户浏览器，客户浏览器发现是 302 响应，则自动再发送一个新的 HTTP 请求，请求的 URL 是新的 location 地址，服务器根据此请求寻找资源并发送给客户。在这里，location 可以重定向到任意 URL。既然是浏览器重新发出了请求，那就没有什么 request 传递的概念了。在客户浏览器的地址栏中显示的是其重定向的路径，客户可以观察到地址的变化。重定向行为是浏览器做了至少两次访问请求。

在 Spring MVC 框架中，控制器类中处理方法的 return 语句默认就是转发实现，只不过实现的是转发到视图。示例代码如下：

```
@RequestMapping("/register")
public String register() {
      return "register";//转发到register.jsp
}
```

在Spring MVC框架中，重定向与转发的示例代码如下：

```
package controller;
import org.springframework.stereotype.Controller;
import org.springframework.web.bind.annotation.RequestMapping;
@Controller
@RequestMapping("/index")
public class IndexController {
    @RequestMapping("/login")
    public String login() {
        //转发到一个请求方法（同一个控制器类里，可省略/index/）
        return "forward:/index/isLogin";
    }
    @RequestMapping("/isLogin")
    public String isLogin() {
        //重定向到一个请求方法
        return "redirect:/index/isRegister";
    }
    @RequestMapping("/isRegister")
    public String isRegister() {
        //转发到一个视图
        return "register";
    }
}
```

在Spring MVC框架中，不管重定向或转发，都需要符合视图解析器的配置。如果直接转发到一个不需要DispatcherServlet的资源，如：

```
return "forward:/html/my.html";
```

那么就需要使用mvc:resources配置：

```
<mvc:resources location="/html/" mapping="/html/**"></mvc:resources>
```

2.4 应用@Autowired和@Service进行依赖注入

在前面学习的控制器中，并没有体现MVC的M层，这是因为控制器既充当C层，又充当M层。这样设计程序的系统结构很不合理，应该将M层从控制器中分离出来。Spring MVC框架本身就是一个非常优秀的MVC框架，它具有一个依赖注入的优点。可以通过 org.springframework.beans.factory.annotation.Autowired 注解类型将依赖注入一个属性（成员变量）或方法，如：

```
@Autowired
public UserService userService;
```

在 Spring MVC 中，为了能被作为依赖注入，类必须使用 org.springframework.stereotype.Service 注解类型注明为@Service（一个服务）。另外，还需要在配置文件中使用<context:component-scan base-package="基本包"/>元素来扫描依赖基本包。下面将第 2.2 节中“登录”和“注册”的业务逻辑处理分离出来，使用 Service 层实现。

首先，创建 service 包，在包中创建 UserService 接口和 UserServiceImpl 实现类。

UserService 接口的具体代码如下：

```
package service;
import domain.UserForm;
public interface UserService {
     boolean login(UserForm user);
     boolean register(UserForm user);
}
```

UserServiceImpl 实现类的具体代码如下：

```
package service;
import org.springframework.stereotype.Service;
import domain.UserForm;
//注解为一个服务
@Service
public class UserServiceImpl implements UserService{
    @Override
    public boolean login(UserForm user) {
        if("zhangsan".equals(user.getUname())
                && "123456".equals(user.getUpass()))
            return true;
        return false;
    }
    @Override
    public boolean register(UserForm user) {
        if("zhangsan".equals(user.getUname())
                && "123456".equals(user.getUpass()))
            return true;
        return false;
    }
}
```

其次，在配置文件中添加一个<context:component-scan base-package="基本包"/>元素，具体代码如下：

```
<context:component-scan base-package="service"/>
```

最后，修改控制器类UserController，具体代码如下：

```
package controller;
import javax.servlet.http.HttpSession;
import org.apache.commons.logging.Log;
import org.apache.commons.logging.LogFactory;
import org.springframework.beans.factory.annotation.Autowired;
import org.springframework.stereotype.Controller;
import org.springframework.ui.Model;
import org.springframework.web.bind.annotation.RequestMapping;
import domain.UserForm;
import service.UserService;
@Controller
@RequestMapping("/user")
public class UserController {
    //得到一个用来记录日志的对象，这样打印信息的时候能够标记打印的是哪个类的信息
    private static final Log logger = LogFactory.getLog(UserController.
class);
    //将服务依赖注入属性userService
    @Autowired
     public UserService userService;
    /**
     * 处理登录
     */
    @RequestMapping("/login")
    public String login(UserForm user, HttpSession session, Model model) {
        if(userService.login(user)){
                session.setAttribute("u", user);
                logger.info("成功");
                return "main";//登录成功，跳转到main.jsp
        }else{
                logger.info("失败");
                model.addAttribute("messageError", "用户名或密码错误");
                return "login";
        }
    }
    /**
     *处理注册
     */
    @RequestMapping("/register")
    public String register(@ModelAttribute("user") UserForm user) {
        if(userService.register(user)){
                logger.info("成功");
                return "login";//注册成功，跳转到login.jsp
```

```
        }else{
            logger.info("失败");
            //使用@ModelAttribute("user")与 model.addAttribute("user",
user)功能相同
           //在 register.jsp 页面上可以使用 EL 表达式${user.uname}取出
ModelAttribute 的 uname 值
            return "register";//返回 register.jsp
        }
    }
}
```

2.5 @ModelAttribute

通过 org.springframework.web.bind.annotation.ModelAttribute 注解类型，通常可实现如下两个功能。

1．绑定请求参数到实体对象（表单的命令对象）

该用法如第 2.2.6 节内容：

```
@RequestMapping("/register")
public String register(@ModelAttribute("user") UserForm user) {
    if("zhangsan".equals(user.getUname())
            && "123456".equals(user.getUpass())){
        return "login";
    }else{
        return "register";
    }
}
```

上述代码中“@ModelAttribute("user") UserForm user”语句的功能有两个，一是将请求参数的输入封装到 user 对象中；一是创建 UserForm 实例，以“user”为键值存储在 Model 对象中，与“model.addAttribute("user", user)”语句功能一样。如果没有指定键值，即“@ModelAttribute UserForm user”，那么创建 UserForm 实例时，会以“userForm”为键值存储在 Model 对象中，与“model.addAttribute("userForm", user)”语句功能一样。

2．注解一个非请求处理方法

被@ModelAttribute 注解的方法，将在每次调用该控制器类的请求处理方法前被调用。这种特性可以用来控制登录权限，当然控制登录权限的方法很多，例如拦截器、过滤器等。

使用该特性控制登录权限的示例代码如下：

```
package controller;
import javax.servlet.http.HttpSession;
import org.springframework.web.bind.annotation.ModelAttribute;
public class BaseController {
    @ModelAttribute
    public void isLogin(HttpSession session) throws Exception {
        if(session.getAttribute("user") == null){
            throw new Exception("没有权限");
        }
    }
}

package controller;
import org.springframework.stereotype.Controller;
import org.springframework.web.bind.annotation.RequestMapping;
@Controller
@RequestMapping("/admin")
public class ModelAttributeController extends BaseController{
    @RequestMapping("/add")
    public String add(){
        return "addSuccess";
    }
    @RequestMapping("/update")
    public String update(){
        return "updateSuccess";
    }
    @RequestMapping("/delete")
    public String delete(){
        return "deleteSuccess";
    }
}
```

上述 ModelAttributeController 类中的 add、update、delete 请求处理方法执行时，首先执行父类 BaseController 中的 isLogin 判断登录权限。可以通过地址 http://localhost:8080/ch2/admin/add 测试登录权限。

2.6 本章小结

本章是整个 Spring MVC 框架的核心部分。通过本章的学习，务必掌握如何编写基于注解的控制器类。

习　题 2

1．在 Spring MVC 的控制器类中如何访问 Servlet API？

2．控制器接收请求参数的常见方式有哪几种？

3．如何编写基于注解的控制器类？

4．@ModelAttribute 可实现哪些功能？

第3章

类型转换和格式化

学习目的与要求

本章主要学习类型转换器和格式化转换器。通过本章的学习，应该理解类型转换器和格式化转换器的原理，掌握类型转换器和格式化转换器的用法。

本章主要内容

- Converter
- Formatter

在 Spring MVC 框架中，需要收集用户请求参数，并将请求参数传递给应用的控制器组件。此时存在一个问题，所有的请求参数类型只能是字符串数据类型，但 Java 是强类型语言，所以 Spring MVC 框架必须将这些字符串请求参数转换成相应的数据类型。

Spring MVC 框架不仅提供了强大的类型转换和格式化机制，而且开发者还可以方便地开发出自己的类型转换器和格式化转换器，完成字符串和各种数据类型之间的转换。这正是学习本章的目的所在。

3.1　类型转换的意义

本节通过一个简单应用（JSP + Servlet）为示例来介绍类型转换的意义。如图 3.1 所示地添加商品页面，该页面用于收集用户输入的商品信息。商品信息包括：商品名称（字符串类型 String）、商品价格（双精度浮点类型 double）、商品数量（整数类型 int）。

http://localhost:8080/ch3a/addGoods.jsp
商品名称：
商品价格：
商品数量：
提交

图 3.1　添加商品信息的收集页面

addGoods.jsp 页面的代码如下：

```
<body>
```

```
    <form action="addGoods" method="post">
         商品名称：<input type="text" name="goodsname"/><br>
         商品价格：<input type="text" name="goodsprice"/><br>
         商品数量：<input type="text" name="goodsnumber"/><br>
         <input type="submit" value="提交"/>
    </form>
</body>
```

希望页面收集到的数据提交到 addGoods 的 Servlet（AddGoodsServlet 类），该 Servlet 将这些请求信息封装成一个 Goods 类的值对象。

Goods 类的代码如下：

```
package domain;
public class Goods {
    private String goodsname;
    private double goodsprice;
    private int goodsnumber;
    //无参数的构造方法
    public Goods(){}
    //有参数的构造方法
    public Goods(String goodsname, double goodsprice, int goodsnumber) {
        super();
        this.goodsname = goodsname;
        this.goodsprice = goodsprice;
        this.goodsnumber = goodsnumber;
    }
    //此处省略了 setter 和 getter 方法
    …
}
```

AddGoodsServlet 类的代码如下：

```
package servlet;
import java.io.IOException;
import javax.servlet.ServletException;
import javax.servlet.http.HttpServlet;
import javax.servlet.http.HttpServletRequest;
import javax.servlet.http.HttpServletResponse;
import domain.Goods;
public class AddGoodsServlet extends HttpServlet {
    public void doGet(HttpServletRequest request, HttpServletResponse
response)
            throws ServletException, IOException {
        doPost(request, response);
    }
    public void doPost(HttpServletRequest request, HttpServletResponse
response)
```

```
        throws ServletException, IOException {
    response.setContentType("text/html;charset=utf-8");
    //设置编码，防止乱码
    request.setCharacterEncoding("utf-8");
    //获取参数值
    String goodsname = request.getParameter("goodsname");
    String goodsprice = request.getParameter("goodsprice");
    String goodsnumber = request.getParameter("goodsnumber");
    //下面进行类型转换
    double newgoodsprice = Double.parseDouble(goodsprice);
    int newgoodsnumber = Integer.parseInt(goodsnumber);
    //将转换后的数据封装成 goods 值对象
    Goods goods = new Goods(goodsname, newgoodsprice, newgoodsnumber);
    //将 goods 值对象传递给数据访问层，进行添加操作，代码省略
    …
  }
}
```

对于上面这个应用而言，开发者需要自己在 Servlet 中进行类型转换，并将其封装成值对象。这些类型转换操作全部手工完成，非常烦琐。

对于 Spring MVC 框架而言，它必须将请求参数转换成值对象类里各属性对应的数据类型——这就是类型转换的意义。

3.2 Converter

Spring MVC 框架的 Converter<S, T>是一个可以将一种数据类型转换成另一种数据类型的接口，这里 S 表示源类型，T 表示目标类型。开发者在实际应用中，使用框架内置的类型转换器基本就够了，但有时需要编写具有特定功能的类型转换器。

3.2.1 内置的类型转换器

在 Spring MVC 框架中，对于常用的数据类型，开发者无须创建自己的类型转换器，因为 Spring MVC 框架有许多内置的类型转换器完成常用的类型转换。Spring MVC 框架提供的内置类型转换包括如下几种类型。

- 标量转换器

StringToBooleanConverter：String 到 boolean 类型转换。

ObjectToStringConverter：Object 到 String 转换，调用 toString 方法转换。

StringToNumberConverterFactory：String 到数字转换（如 Integer、Long 等）。

NumberToNumberConverterFactory：数字子类型（基本类型）到数字类型（包装类型）转换。

StringToCharacterConverter：String 到 Character 转换，取字符串第一个字符。

NumberToCharacterConverter：数字子类型到 Character 转换。

CharacterToNumberFactory：Character 到数字子类型转换。

StringToEnumConverterFactory：String 到枚举类型转换，通过 Enum.valueOf 将字符串转换为需要的枚举类型。

EnumToStringConverter：枚举类型到 String 转换，返回枚举对象的 name()值。

StringToLocaleConverter：String 到 java.util.Locale 转换。

PropertiesToStringConverter：java.util.Properties 到 String 转换，默认通过 ISO-8859-1 解码。

StringToPropertiesConverter：String 到 java.util.Properties 转换，默认使用 ISO-8859-1 编码。

- 集合、数组相关转换器

ArrayToCollectionConverter：任意数组到任意集合（List、Set）转换。

CollectionToArrayConverter：任意集合到任意数组转换。

ArrayToArrayConverter：任意数组到任意数组转换。

CollectionToCollectionConverter：集合之间的类型转换。

MapToMapConverter：Map 之间的类型转换。

ArrayToStringConverter：任意数组到 String 转换。

StringToArrayConverter：字符串到数组的转换，默认通过“,”分隔，且去除字符串的两边空格（trim）。

ArrayToObjectConverter：任意数组到 Object 的转换，如果目标类型和源类型兼容，直接返回源对象；否则返回数组的第一个元素并进行类型转换。

ObjectToArrayConverter：Object 到单元素数组转换。

CollectionToStringConverter：任意集合（List、Set）到 String 转换。

StringToCollectionConverter：String 到集合（List、Set）转换，默认通过“,”分隔，且去除字符串的两边空格（trim）。

CollectionToObjectConverter：任意集合到任意 Object 的转换，如果目标类型和源类型兼容，直接返回源对象；否则返回集合的第一个元素并进行类型转换。

ObjectToCollectionConverter：Object 到单元素集合的类型转换。

类型转换是在视图与控制器相互传递数据时发生的。Spring MVC 框架对于基本类型如 int、long、float、double、boolean 以及 char 等，已经做好了基本类型转换。例如，针对第 3.1 节 addGoods.jsp 的提交请求，可以由如下处理方法来接收请求参数并处理：

```
package controller;
import org.springframework.stereotype.Controller;
import org.springframework.web.bind.annotation.RequestMapping;
@Controller
public class GoodsController {
    @RequestMapping("/addGoods")
    public String add(String goodsname, double goodsprice, int goodsnumber){
        double total = goodsprice * goodsnumber;
```

```
            System.out.println(total);
            return "success";
        }
    }
```

注意：内置类型转换器使用时，请求参数输入值与接收参数类型要兼容，否则报400错误。请求参数类型与接收参数类型不兼容问题，需要学习输入校验后才可解决。

3.2.2 自定义类型转换器

当 Spring MVC 框架内置的类型转换器不能满足需求时，开发者可以开发自己的类型转换器。例如有个应用 ch3a 希望用户在页面表单中输入信息来创建商品信息。当输入“apple,10.58,200”时，表示在程序中自动创建一个 new Goods，并将“apple”值自动赋值给 goodsname 属性，将“10.58”值自动赋值给 goodsprice 属性，将“200”值自动赋值给 goodsnumber 属性。ch3a 应用的项目结构如图 3.2 所示。

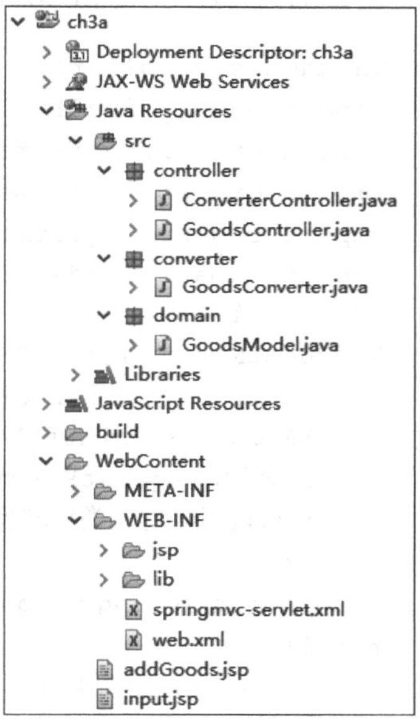

```
ch3a
  Deployment Descriptor: ch3a
  JAX-WS Web Services
  Java Resources
    src
      controller
        ConverterController.java
        GoodsController.java
      converter
        GoodsConverter.java
      domain
        GoodsModel.java
    Libraries
  JavaScript Resources
  build
  WebContent
    META-INF
    WEB-INF
      jsp
      lib
      springmvc-servlet.xml
      web.xml
    addGoods.jsp
    input.jsp
```

图 3.2　ch3a 的项目结构

想实现上述应用需要做以下 5 件事：

（1）创建实体类

（2）创建控制器类

（3）创建自定义类型转换器

（4）注册类型转换器

（5）创建相关视图

按照上述步骤采用自定义类型转换器完成需求。

第 1 步　创建实体类

创建名为 GoodsModel.java 的类文件，代码如下：

```
package domain;
public class GoodsModel {
    private String goodsname;
    private double goodsprice;
    private int goodsnumber;
    public String getGoodsname() {
        return goodsname;
    }
    public void setGoodsname(String goodsname) {
        this.goodsname = goodsname;
    }
    public double getGoodsprice() {
        return goodsprice;
    }
```

```
    public void setGoodsprice(double goodsprice) {
        this.goodsprice = goodsprice;
    }
    public int getGoodsnumber() {
        return goodsnumber;
    }
    public void setGoodsnumber(int goodsnumber) {
        this.goodsnumber = goodsnumber;
    }
}
```

第 2 步　创建控制器类

创建名为 ConverterController.java 的类文件，代码如下：

```
package controller;
import org.springframework.stereotype.Controller;
import org.springframework.ui.Model;
import org.springframework.web.bind.annotation.RequestMapping;
import org.springframework.web.bind.annotation.RequestParam;
import domain.GoodsModel;
@Controller
@RequestMapping("/my")
public class ConverterController {
    @RequestMapping("/converter")
    /*使用@RequestParam("goods")接收请求参数,
    然后调用自定义类型转换器 GoodsConverter 将字符串值转换为 GoodsModel 的对象 gm
    */
    public String myConverter(@RequestParam("goods") GoodsModel gm, Model
model){
        model.addAttribute("goods",gm);
        return "showGoods";
    }
}
```

第 3 步　创建自定义类型转换器

自定义类型转换器类需要实现 Converter<S, T>接口，重写 convert(S)接口方法。convert(S)方法功能是将源数据类型 S 转换成目标数据类型 T。自定义类型转换器类 GoodsConverter 的代码如下：

```
package converter;
import org.springframework.core.convert.converter.Converter;
import domain.GoodsModel;
public class GoodsConverter implements Converter<String, GoodsModel>{
    @Override
    public GoodsModel convert(String source) {
        //创建一个 Goods 实例
```

```
        GoodsModel goods = new GoodsModel();
        //以“,”分隔的
        String stringValues[] = source.split(",");
        if(stringValues != null &&
                stringValues.length == 3){
            //为 Goods 实例赋值
            goods.setGoodsname(stringValues[0]);
            goods.setGoodsprice(Double.parseDouble(stringValues[1]));
            goods.setGoodsnumber(Integer.parseInt(stringValues[2]));
            return goods;
        }else{
            throw new IllegalArgumentException(String.format("类型转换失
败，需要格式‘apple,10.58,200’，但格式是[%s]", source));
        }
    }
}
```

第 4 步　注册类型转换器

在配置文件 springmvc-servlet.xml 中注册自定义类型转换器，配置文件代码如下：

```
<?xml version="1.0" encoding="UTF-8"?>
<beans xmlns="http://www.springframework.org/schema/beans"
   xmlns:xsi="http://www.w3.org/2001/XMLSchema-instance"
   xmlns:p="http://www.springframework.org/schema/p"
   xmlns:context="http://www.springframework.org/schema/context"
   xmlns:mvc="http://www.springframework.org/schema/mvc"
   xsi:schemaLocation="
    http://www.springframework.org/schema/beans
    http://www.springframework.org/schema/beans/spring-beans.xsd
       http://www.springframework.org/schema/context
       http://www.springframework.org/schema/context/
spring-context.xsd
       http://www.springframework.org/schema/mvc
       http://www.springframework.org/schema/mvc/spring-mvc.xsd">
   <!-- 使用扫描机制，扫描 controller 包 -->
   <context:component-scan base-package="controller"/>
   <!-- 注册类型转换器 GoodsConverter -->
<bean id="conversionService" class="org.springframework.context.
support.ConversionServiceFactoryBean">
        <property name="converters">
            <list>
                <bean class="converter.GoodsConverter"/>
            </list>
        </property>
   </bean>
   <mvc:annotation-driven conversion-service="conversionService"/>
```

```
    <!-- 配置视图解析器 -->
    <bean class="org.springframework.web.servlet.view.
InternalResourceViewResolver"
          id="internalResourceViewResolver">
      <!-- 前缀 -->
      <property name="prefix" value="/WEB-INF/jsp/" />
      <!-- 后缀 -->
      <property name="suffix" value=".jsp" />
  </bean>
</beans>
```

第 5 步　创建相关视图

信息采集页面 input.jsp 的代码如下：

```
<%@ page language="java" contentType="text/html; charset=UTF-8"
pageEncoding="UTF-8"%>
<%
String path = request.getContextPath();
String basePath = request.getScheme()+"://"+request.getServerName()+":"+
request.getServerPort()+path+"/";
%>
<!DOCTYPE html PUBLIC "-//W3C//DTD HTML 4.01 Transitional//EN"
"http://www.w3.org/TR/html4/loose.dtd">
<html>
<head>
<base href="<%=basePath%>">
<meta http-equiv="Content-Type" content="text/html; charset=UTF-8">
<title>Insert title here</title>
</head>
<body>
    <form action="my/converter" method="post">
        请输入商品信息（格式为：apple,10.58,200）：
        <input type="text" name="goods"/><br>
        <input type="submit" value="提交"/>
    </form>
</body>
</html>
```

信息显示页面 showGoods.jsp 的代码如下：

```
<%@  page  language="java"  contentType="text/html;  charset=UTF-8"
pageEncoding="UTF-8"%>
<%
String path = request.getContextPath();
String basePath = request.getScheme()+"://"+request.getServerName()+":"+
request.getServerPort()+path+"/";
%>
```

```
<!DOCTYPE html PUBLIC "-//W3C//DTD HTML 4.01 Transitional//EN"
"http://www.w3.org/TR/html4/loose.dtd">
<html>
<head>
<base href="<%=basePath%>">
<meta http-equiv="Content-Type" content="text/html; charset=UTF-8">
<title>Insert title here</title>
</head>
<body>
    您创建的商品信息如下：<br>
  <!-- 使用EL表达式取出model中goods的信息 -->
    商品名为：${goods.goodsname },
    商品价格为：${goods.goodsprice },
    商品数量为：${goods.goodsnumber }。
</body>
</html>
```

最后，使用地址 http://localhost:8080/ch3a/input.jsp 测试应用。

3.2.3 实践环节

创建一个 Web 应用 project323pratice，该应用具体实施步骤如下：

（1）编写一个 JSP 页面 input.jsp，该页面运行效果如图 3.3 所示。

（2）编写实体类 User。

（3）编写控制器类，在控制器类中，类型转换器自动将请求过来的值转换成 User 类型。

（4）编写自定义类型转换器类 UserConverter。

（5）注册类型转换器。

（6）编写用户信息输出页面 showUser.jsp，页面效果如图 3.4 所示。

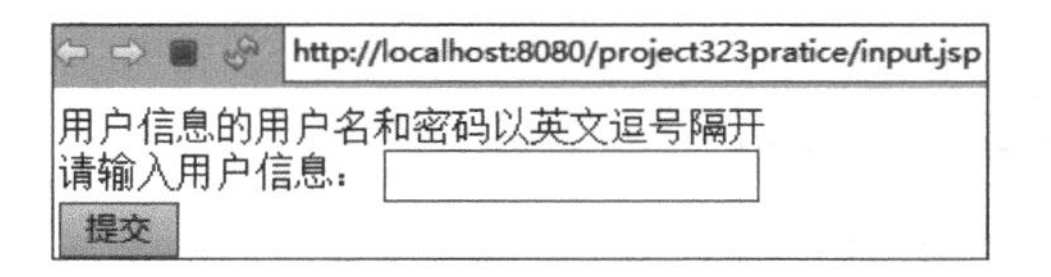

图 3.3 实践环节的首页面

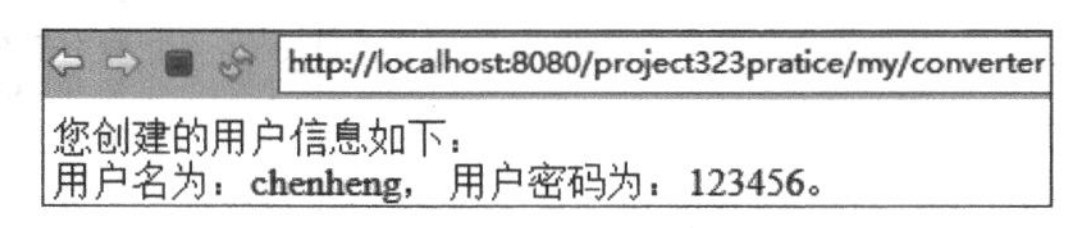

图 3.4 实践环节的结果页面

3.3 Formatter

Spring MVC 框架的 Formatter<T>与 Converter<S, T>一样，也是一个可以将一种数据类型转换成另一种数据类型的接口。但不同的是，Formatter<T>的源数据类型必须是 String 类型，而 Converter<S, T>的源数据类型是任意数据类型。

在 Web 应用中，由 HTTP 发送的请求数据到控制器中都是以 String 类型获取。因此，在 Web 应用中选择 Formatter<T>比选择 Converter<S, T>更加合理。

3.3.1 内置的格式化转换器

Spring MVC 提供几个内置的格式化转换器，具体如下。

- NumberFormatter：实现 Number 与 String 之间的解析与格式化。
- CurrencyFormatter：实现 Number 与 String 之间的解析与格式化（带货币符号）。
- PercentFormatter：实现 Number 与 String 之间的解析与格式化（带百分数符号）。
- DateFormatter：实现 Date 与 String 之间的解析与格式化。

3.3.2 自定义格式化转换器

自定义格式化转换器，就是编写一个实现 org.springframework.format.Formatter 接口的 Java 类。该接口声明如下：

```
public interface Formatter<T>
```

这里的 T 表示由字符串转换的目标数据类型。该接口有 parse 和 print 两个接口方法，自定义格式化转换器类必须覆盖它们。

```
public T parse(String s, java.util.Locale locale)
public String print(T object, java.util.Locale locale)
```

parse 方法的功能是利用指定的 Locale 将一个 String 类型转换成目标类型，print 方法与之相反，返回目标对象的字符串表示。

下面通过具体应用 ch3b 讲解自定义格式化转换器的用法。ch3b 的项目结构如图 3.5 所示。

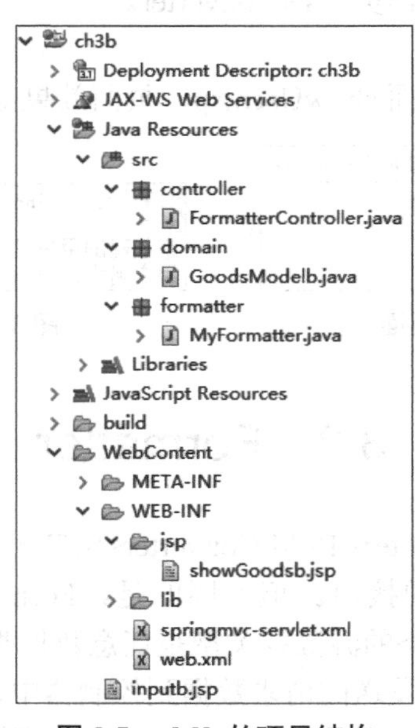

图 3.5 ch3b 的项目结构

应用的具体要求如下：

（1）用户在页面表单中输入信息来创建商品，输入页面效果如图 3.6 所示。

（2）控制器使用实体 bean 类 GoodsModelb 接收页面提交的请求参数，GoodsModelb 类的属性有：

```
private String goodsname;
private double goodsprice;
private int goodsnumber;
private Date goodsdate;
```

（3）GoodsModelb 实体类接收请求参数时，商品名称、价格和数量使用内置的类型转换器完成转换；商品日期需要自定义的格式化转换器完成。

（4）格式化转换器转换之后的数据显示在 showGoodsb.jsp 页面，效果如图 3.7 所示。

http://localhost:8080/ch3b/inputb.jsp
商品名称：aaa
商品价格：12
商品数量：34
商品日期：2017-02-05 （yyyy-MM-dd）
提交

图 3.6 信息输入页面

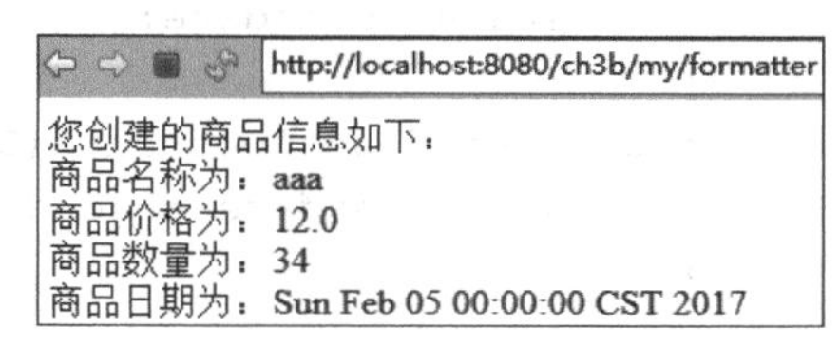

图 3.7 格式化后信息显示页面

由图 3.7 可看出，日期由字符串值“2017-02-05”格式化成 Date 类型。

想实现上述应用 ch3b 的需求，需要做以下 5 件事：

（1）创建实体类

（2）创建控制器类

（3）创建自定义格式化转换器

（4）注册格式化转换器

（5）创建相关视图

按照上述步骤采用自定义格式化转换器完成需求。

第 1 步　创建实体类

实体类 GoodsModelb.java 的代码如下：

```
package domain;
import java.util.Date;
public class GoodsModelb {
    private String goodsname;
    private double goodsprice;
    private int goodsnumber;
    private Date goodsdate;
    public String getGoodsname() {
        return goodsname;
    }
    public void setGoodsname(String goodsname) {
        this.goodsname = goodsname;
```

```
    }
    public double getGoodsprice() {
        return goodsprice;
    }
    public void setGoodsprice(double goodsprice) {
        this.goodsprice = goodsprice;
    }
    public int getGoodsnumber() {
        return goodsnumber;
    }
    public void setGoodsnumber(int goodsnumber) {
        this.goodsnumber = goodsnumber;
    }
    public Date getGoodsdate() {
        return goodsdate;
    }
    public void setGoodsdate(Date goodsdate) {
        this.goodsdate = goodsdate;
    }
}
```

第 2 步　创建控制器类

控制器类 FormatterController.java 的代码如下：

```
package controller;
import org.springframework.stereotype.Controller;
import org.springframework.ui.Model;
import org.springframework.web.bind.annotation.RequestMapping;
import domain.GoodsModelb;
@Controller
@RequestMapping("/my")
public class FormatterController {
    @RequestMapping("/formatter")
    public String myConverter(GoodsModelb gm, Model model){
        model.addAttribute("goods",gm);
        return "showGoodsb";
    }
}
```

第 3 步　创建格式化转换器类

自定义格式化转换器类 MyFormatter.java 的代码如下：

```
package formatter;
import java.text.ParseException;
import java.text.SimpleDateFormat;
import java.util.Date;
import java.util.Locale;
```

```
import org.springframework.format.Formatter;
public class MyFormatter implements Formatter<Date>{
    SimpleDateFormat dateFormat = new SimpleDateFormat("yyyy-MM-dd");
    @Override
    public String print(Date object, Locale arg1) {
        return dateFormat.format(object);
    }
    @Override
    public Date parse(String source, Locale arg1) throws ParseException {
        return dateFormat.parse(source);//Formatter只能对字符串转换
    }
}
```

第4步 注册格式化转换器

在配置文件中注册格式化转换器，具体代码如下：

```
<?xml version="1.0" encoding="UTF-8"?>
<beans xmlns="http://www.springframework.org/schema/beans"
   xmlns:xsi="http://www.w3.org/2001/XMLSchema-instance"
   xmlns:p="http://www.springframework.org/schema/p"
   xmlns:context="http://www.springframework.org/schema/context"
   xmlns:mvc="http://www.springframework.org/schema/mvc"
   xsi:schemaLocation="
    http://www.springframework.org/schema/beans
    http://www.springframework.org/schema/beans/spring-beans.xsd
       http://www.springframework.org/schema/context
       http://www.springframework.org/schema/context/spring-context.xsd
       http://www.springframework.org/schema/mvc
       http://www.springframework.org/schema/mvc/spring-mvc.xsd">
   <!-- 使用扫描机制，扫描controller包 -->
   <context:component-scan base-package="controller"/>
   <!-- 注册MyFormatter-->
       <bean id="conversionService" class="org.springframework.format.
support.FormattingConversionServiceFactoryBean">
    <property name="formatters">
        <set>
            <bean class="formatter.MyFormatter"/>
        </set>
    </property>
   </bean>
   <mvc:annotation-driven conversion-service="conversionService"/>
    <!-- 配置视图解析器 -->
    <bean class="org.springframework.web.servlet.view.
InternalResourceViewResolver"
           id="internalResourceViewResolver">
      <!-- 前缀 -->
```

```
        <property name="prefix" value="/WEB-INF/jsp/" />
        <!-- 后缀 -->
        <property name="suffix" value=".jsp" />
    </bean>
</beans>
```

第5步　创建相关视图

信息输入页面 inputb.jsp 的代码如下：

```
<%@ page language="java" contentType="text/html; charset=UTF-8"
pageEncoding="UTF-8"%>
<%
String path = request.getContextPath();
String basePath = request.getScheme()+"://"+request.getServerName()+":"+
request.getServerPort()+path+"/";
%>
<!DOCTYPE html PUBLIC "-//W3C//DTD HTML 4.01 Transitional//EN"
"http://www.w3.org/TR/html4/loose.dtd">
<html>
<head>
<base href="<%=basePath%>">
<meta http-equiv="Content-Type" content="text/html; charset=UTF-8">
<title>Insert title here</title>
</head>
<body>
    <form action="my/formatter" method="post">
        商品名称：<input type="text" name="goodsname"/><br>
        商品价格：<input type="text" name="goodsprice"/><br>
        商品数量：<input type="text" name="goodsnumber"/><br>
        商品日期：<input type="text" name="goodsdate"/>（yyyy-MM-dd）<br>
        <input type="submit" value="提交"/>
    </form>
</body>
</html>
```

信息显示页面 showGoodsb.jsp 的代码如下：

```
<%@ page language="java" contentType="text/html; charset=UTF-8"
pageEncoding="UTF-8"%>
<%
String path = request.getContextPath();
String basePath = request.getScheme()+"://"+request.getServerName()+":"+
request.getServerPort()+path+"/";
%>
<!DOCTYPE html PUBLIC "-//W3C//DTD HTML 4.01 Transitional//EN"
"http://www.w3.org/TR/html4/loose.dtd">
<html>
```

```
<head>
<base href="<%=basePath%>">
<meta http-equiv="Content-Type" content="text/html; charset=UTF-8">
<title>Insert title here</title>
</head>
<body>
    您创建的商品信息如下：<br>
  <!-- 使用EL表达式取出Action类的属性goods的值 -->
    商品名称为：${goods.goodsname }<br>
    商品价格为：${goods.goodsprice }<br>
    商品数量为：${goods.goodsnumber }<br>
    商品日期为：${goods.goodsdate }
</body>
</html>
```

最后，通过地址 http://localhost:8080/ch3b/inputb.jsp 测试应用。

3.3.3 实践环节

创建一个 Web 应用 project333pratice，在该应用中将第 3.2.3 节 的自定义类型转换器修改成自定义格式化转换器。

3.4 本章小结

本章重点讲解了自定义类型转换器和格式化转换器的实现和注册。但在实际应用中，开发者很少自定义类型转换器和格式化类型转换器，一般都是使用内置的转换器。

习题 3

1．在 MVC 框架中，为什么要进行类型转换？

2．Converter 与 Formatter 的区别是什么？

3．在 Spring MVC 框架中，如何自定义类型转换器类，又如何注册类型转换器？

4．在 Spring MVC 框架中，如何自定义格式化转换器类，又如何注册格式化转换器？

第4章 chapter 4

数据绑定和表单标签库

学习目的与要求

本章主要讲解数据绑定和表单标签库。通过本章的学习，理解数据绑定的基本原理，掌握表单标签库的用法。

本章主要内容

- 数据绑定
- 表单标签库
- 数据绑定应用

数据绑定是将用户参数输入值绑定到领域模型的一种特性，在 Spring MVC 的 Controller 和 View 参数数据传递中，所有 HTTP 请求参数的类型均为字符串，如果模型需要绑定的类型为 double 或 int，则需要手动进行类型转换，而有了数据绑定后，就不再需要手动将 HTTP 请求中的 String 类型转换为模型需要的类型。数据绑定的另一个好处是，当输入验证失败时，会重新生成一个 HTML 表单，无须重新填写输入字段。

在 Spring MVC 中，为了方便、高效地使用数据绑定，还需要学习表单标签库。

4.1 数据绑定

在 Spring MVC 框架中，数据绑定有这样几层含义：绑定请求参数输入值到领域模型（如第 2.2 节）、模型数据到视图的绑定（输入验证失败时）、模型数据到表单元素的绑定（如下列列表选项值由控制器初始化）。有关数据绑定的示例，请参见第 4.3 节数据绑定应用。

4.2 表单标签库

表单标签库中包含了可以用在 JSP 页面中渲染 HTML 元素的标签。JSP 页面使用 Spring 表单标签库时，必须在 JSP 页面开头处声明 taglib 指令，指令代码如下：

```
<%@ taglib prefix="form" uri="http://www.springframework.org/tags/form" %>
```

表单标签库中有 form、input、password、hidden、textarea、checkbox、checkboxes、radiobutton、radiobuttons、select、option、options、errors。

form：渲染表单元素。

input：渲染<input type="text"/>元素。

password：渲染<input type="password"/>元素。

hidden：渲染<input type="hidden"/>元素。

textarea：渲染 textarea 元素。

checkbox：渲染一个<input type="checkbox"/>元素。

checkboxes：渲染多个<input type="checkbox"/>元素。

radiobutton：渲染一个<input type="radio"/>元素。

radiobuttons：渲染多个<input type="radio"/>元素。

select：渲染一个选择元素。

option：渲染一个选项元素。

options：渲染多个选项元素。

errors：在 span 元素中渲染字段错误。

4.2.1 表单标签

表单标签，语法格式如下：

```
<form:form modelAttribute="xxx" method="post" action="xxx">
      ...
</form:form>
```

除了具有 HTML 表单元素属性外，表单标签还具有 acceptCharset、commandName、cssClass、cssStyle、htmlEscape 和 modelAttribute 等属性。各属性含义如下所示。

acceptCharset：定义服务器接受的字符编码列表。

commandName：暴露表单对象的模型属性名称，默认为 command。

cssClass：定义应用到 form 元素的 CSS 类。

cssStyle：定义应用到 form 元素的 CSS 样式。

htmlEscape：true 或 false，表示是否进行 HTML 转义。

modelAttribute：暴露 form backing object 的模型属性名称，默认为 command。

其中，commandName 和 modelAttribute 属性功能基本一致，属性值绑定一个 JavaBean 对象。假设，控制器类 UserController 的方法 inputUser 是返回 userAdd.jsp 的请求处理方法。

inputUser 方法的代码如下：

```
@RequestMapping(value = "/input")
public String inputUser(Model model) {
```

```
    ...
    model.addAttribute("user", new User());
    return "userAdd";
}
```

userAdd.jsp 的表单标签代码如下：

```
<form:form modelAttribute="user"  method="post" action="user/save">
    ...
</form:form>
```

注意：在 inputUser 方法中，如果没有 Model 属性 user，userAdd.jsp 页面就会抛出异常，因为表单标签无法找到在其 modelAttribute 属性中指定的 form backing object。

4.2.2 input 标签

input 标签语法格式如下：

```
<form:input path="xxx"/>
```

该标签除了 cssClass、cssStyle、htmlEscape 属性外，还有一个最重要的属性 path。path 属性将文本框输入值绑定到 form backing object 的一个属性。示例代码如下：

```
<form:form modelAttribute="user"  method="post" action="user/save">
    <form:input path="userName"/>
</form:form>
```

上述代码将输入值绑定到 user 对象的 userName 属性。

4.2.3 password 标签

password 标签语法格式如下：

```
<form:password path="xxx"/>
```

该标签与 input 标签用法完全一致，这里不再赘述。

4.2.4 hidden 标签

hidden 标签语法格式如下：

```
<form:hidden path="xxx"/>
```

该标签与 input 标签用法基本一致，只不过它不可显示，不支持 cssClass 和 cssStyle 属性。

4.2.5 textarea 标签

textarea 基本上就是一个支持多行输入的 input 元素，语法格式如下：

```
<form:textarea path="xxx"/>
```

该标签与 input 标签用法完全一致，这里不再赘述。

4.2.6 checkbox 标签

checkbox 标签语法格式如下：

```
<form:checkbox path="xxx" value="xxx"/>
```

多个 path 相同的 checkbox 标签，它们是一个选项组，允许多选。选项值绑定到一个数组属性。示例代码如下：

```
<form:checkbox path="friends" value="张三"/>张三
<form:checkbox path="friends" value="李四"/>李四
<form:checkbox path="friends" value="王五"/>王五
<form:checkbox path="friends" value="赵六"/>赵六
```

上述示例代码中复选框的值绑定到一个字符串数组属性 friends（String[] friends）。该标签的其他用法与 input 标签基本一致，这里不再赘述。

4.2.7 checkboxes 标签

checkboxes 标签渲染多个复选框，是一个选项组，等价于多个 path 相同的 checkbox 标签。它有三个非常重要的属性：items、itemLabel 和 itemValue。

items：用于生成 input 元素的 Collection、Map 或 Array。

itemLabel：items 属性中指定的集合对象的属性，为每个 input 元素提供 label。

itemValue：items 属性中指定的集合对象的属性，为每个 input 元素提供 value。

checkboxes 标签语法格式如下：

```
<form:checkboxes items="xxx"  path="xxx"/>
```

示例代码如下：

```
<form:checkboxes items="${hobbys}"  path="hobby" />
```

上述示例代码是将 model 属性 hobbys 的内容（集合元素）渲染为复选框。itemLabel 和 itemValue 默认情况下，如果集合是数组，复选框的 label 和 value 相同；如果是 Map 集合，复选框的 label 是 Map 的值（value），复选框的 value 是 Map 的关键字（key）。

4.2.8 radiobutton 标签

radiobutton 标签语法格式如下：

```
<form:radiobutton path="xxx" value="xxx"/>
```

多个 path 相同的 radiobutton 标签，它们是一个选项组，只允许单选。

4.2.9 radiobuttons 标签

radiobuttons 标签渲染多个 radio，是一个选项组，等价于多个 path 相同的 radiobutton 标签。radiobuttons 标签语法格式如下：

```
<form:radiobuttons path="xxx" items="xxx"/>
```

该标签的 itemLabel 和 itemValue 属性与 checkboxes 标签的 itemLabel 和 itemValue 属性完全一样，但只允许单选。

4.2.10 select 标签

select 标签的选项可能来自其属性 items 指定的集合，或者来自一个嵌套的 option 标签或 options 标签。语法格式如下：

```
<form:select path="xxx" items="xxx" />
```

或

```
<form:select path="xxx" items="xxx" >
        <option value="xxx">xxx</option>
</form:select>
```

或

```
<form:select path="xxx">
     <form:options items="xxx"/>
</form:select>
```

该标签的 itemLabel 和 itemValue 属性与 checkboxes 标签的 itemLabel 和 itemValue 属性完全一样。

4.2.11 options 标签

options 标签生成一个 select 标签的选项列表。因此，需要与 select 标签一同使用，具体用法参见第 4.2.10 节 select 标签。

4.2.12 errors 标签

errors 标签渲染一个或者多个 span 元素，每个 span 元素包含一个错误消息。它可以用于显示一个特定的错误消息，也可以显示所有错误消息。语法如下：

```
<form:errors path="*"/>
```

或

```
<form:errors path="xxx"/>
```

其中，“*”表示显示所有错误消息；“xxx”表示显示由“xxx”指定的特定错误消息。

4.3 数据绑定应用

为了让读者进一步学习数据绑定和表单标签，本节给出了一个应用范例 ch4。应用中实现了 User 类属性和 JSP 页面中表单参数的绑定，同时在 JSP 页面中分别展示了 input、textarea、checkbox、checkboxs、select 等标签。

4.3.1 应用的相关配置

在 ch4 应用中需要使用 JSTL，因此，不仅需要将 Spring MVC 相关 jar 包复制到应用的 WEN-INF/lib 目录下，还需要从 Tomcat 的 webapps\examples\WEB-INF\lib 目录下，将 JSTL 相关 jar 包复制到应用的 WEN-INF/lib 目录下。ch4 的项目结构如图 4.1 所示。

图 4.1 ch4 的项目结构

为了避免中文乱码问题，需要在 web.xml 文件中增加编码过滤器，同时 JSP 页面编码设置为 UTF-8，form 表单的提交方式必须为 post。

web.xml 的代码如下：

```
<?xml version="1.0" encoding="UTF-8"?>
<web-app xmlns:xsi="http://www.w3.org/2001/XMLSchema-instance"
    xmlns="http://xmlns.jcp.org/xml/ns/javaee"
    xsi:schemaLocation="http://xmlns.jcp.org/xml/ns/javaee
http://xmlns.jcp.org/xml/ns/javaee/web-app_3_1.xsd"
    id="WebApp_ID" version="3.1">
    <display-name>cardManage</display-name>
    <welcome-file-list>
        <welcome-file>index.html</welcome-file>
        <welcome-file>index.htm</welcome-file>
        <welcome-file>index.jsp</welcome-file>
        <welcome-file>default.html</welcome-file>
        <welcome-file>default.htm</welcome-file>
        <welcome-file>default.jsp</welcome-file>
    </welcome-file-list>
    <!--配置 DispatcherServlet -->
    <servlet>
        <servlet-name>springmvc</servlet-name>
    <servlet-class>org.springframework.web.servlet.DispatcherServlet
</servlet-class>
        <load-on-startup>1</load-on-startup>
    </servlet>
    <servlet-mapping>
        <servlet-name>springmvc</servlet-name>
        <url-pattern>/</url-pattern>
    </servlet-mapping>
    <!-- 避免中文乱码 -->
    <filter>
        <filter-name>characterEncodingFilter</filter-name>
    <filter-class>org.springframework.web.filter.
CharacterEncodingFilter</filter-class>
        <init-param>
            <param-name>encoding</param-name>
            <param-value>UTF-8</param-value>
        </init-param>
        <init-param>
            <param-name>forceEncoding</param-name>
            <param-value>true</param-value>
        </init-param>
    </filter>
    <filter-mapping>
        <filter-name>characterEncodingFilter</filter-name>
```

```
        <url-pattern>/*</url-pattern>
    </filter-mapping>
</web-app>
```

配置文件 springmvc-servlet.xml 与前文学习过的配置文件没有区别，这里不再赘述。

4.3.2 领域模型

应用中实现了 User 类属性和 JSP 页面中表单参数的绑定，User 类包含了和表单参数名对应的属性，以及属性的 set 和 get 方法。

User.java 的代码如下：

```
package domain;
public class User {
    private String userName;
    private String[] hobby;//兴趣爱好
    private String[] friends;//朋友
    private String carrer;
    private String houseRegister;
    private String remark;
    public String getUserName() {
        return userName;
    }
    public void setUserName(String userName) {
        this.userName = userName;
    }
    public String[] getHobby() {
        return hobby;
    }
    public void setHobby(String[] hobby) {
        this.hobby = hobby;
    }
    public String[] getFriends() {
        return friends;
    }
    public void setFriends(String[] friends) {
        this.friends = friends;
    }
    public String getCarrer() {
        return carrer;
    }
    public void setCarrer(String carrer) {
        this.carrer = carrer;
    }
    public String getHouseRegister() {
```

```
        return houseRegister;
    }
    public void setHouseRegister(String houseRegister) {
        this.houseRegister = houseRegister;
    }
    public String getRemark() {
        return remark;
    }
    public void setRemark(String remark) {
        this.remark = remark;
    }
}
```

4.3.3 Service 层

应用中使用了 Service 层，在 Service 层使用静态集合变量 users 模拟数据库存储用户信息，包括添加用户和查询用户两个功能方法。

UserService.java 的代码如下：

```
package service;
import java.util.ArrayList;
import domain.User;
public interface UserService {
    boolean addUser(User u);
    ArrayList<User> getUsers();
}
```

UserServiceImpl.java 的代码如下：

```
package service;
import java.util.ArrayList;
import org.springframework.stereotype.Service;
import domain.User;
@Service
public class UserServiceImpl implements UserService{
    //使用静态集合变量 users 模拟数据库
    private static ArrayList<User> users = new ArrayList<User>();
    @Override
    public boolean addUser(User u) {
        if(!"IT 民工".equals(u.getCarrer())){//不允许添加 IT 民工
            users.add(u);
            return true;
        }
        return false;
    }
    @Override
```

```
    public ArrayList<User> getUsers() {
        return users;
    }
}
```

4.3.4 Controller 层

在 Controller 类 UserController 中定义了请求处理方法，其中包括处理 user/input 请求的 inputUser 方法，以及 user/save 请求的 addUser 方法，其中在 addUser 方法中用到了重定向。在 UserController 类中，通过@Autowired 注解在 UserController 对象中主动注入 UserService 对象，实现对 user 对象的添加和查询等操作；通过 model 的 addAttribute 方法将 User 类对象、HashMap 类型的 hobbys 对象、String[]类型的 carrers 对象以及 String[]类型的 houseRegisters 对象传递给 View（userAdd.jsp）。

UserController.java 的代码如下：

```
package controller;
import java.util.HashMap;
import java.util.List;
import org.apache.commons.logging.Log;
import org.apache.commons.logging.LogFactory;
import org.springframework.beans.factory.annotation.Autowired;
import org.springframework.stereotype.Controller;
import org.springframework.ui.Model;
import org.springframework.web.bind.annotation.ModelAttribute;
import org.springframework.web.bind.annotation.RequestMapping;
import domain.User;
import service.UserService;
@Controller
@RequestMapping("/user")
public class UserController {
    // 得到一个用来记录日志的对象，这样打印信息的时候能够标记打印的是哪个类的信息
    private static final Log logger = LogFactory.getLog(UserController.
class);
    @Autowired
    private UserService userService;
    @RequestMapping(value = "/input")
    public String inputUser(Model model) {
        HashMap<String, String> hobbys = new HashMap<String, String>();
        hobbys.put("篮球", "篮球");
        hobbys.put("乒乓球", "乒乓球");
        hobbys.put("电玩", "电玩");
        hobbys.put("游泳", "游泳");
        //如果 model 中没有 user 属性，userAdd.jsp 会抛出异常，因为表单标签无法找到
// modelAttribute 属性指定的 form backing object
```

```
        model.addAttribute("user", new User());
        model.addAttribute("hobbys", hobbys);
    model.addAttribute("carrers", new String[] { "教师", "学生", "coding
搬运工", "IT 民工", "其他" });
        model.addAttribute("houseRegisters", new String[] { "北京", "上
海", "广州", "深圳", "其他" });
        return "userAdd";
    }
    @RequestMapping(value = "/save")
    public String addUser(@ModelAttribute User user, Model model) {
        if (userService.addUser(user)) {
            logger.info("成功");
            return "redirect:/user/list";
        } else {
            logger.info("失败");
            HashMap<String,  String>  hobbys  =  new  HashMap<String,
String>();
            hobbys.put("篮球", "篮球");
            hobbys.put("乒乓球", "乒乓球");
            hobbys.put("电玩", "电玩");
            hobbys.put("游泳", "游泳");
            // 这里不需要 model.addAttribute("user", new
            // User()), 因为@ModelAttribute 指定 form backing object
            model.addAttribute("hobbys", hobbys);
    model.addAttribute("carrers", new String[] { "教师", "学生", "coding
搬运工", "IT 民工", "其他" });
        model.addAttribute("houseRegisters", new String[] { "北京", "上
海", "广州", "深圳", "其他" });
            return "userAdd";
        }
    }
    @RequestMapping(value = "/list")
    public String listUsers(Model model) {
        List<User> users = userService.getUsers();
        model.addAttribute("users", users);
        return "userList";
    }
}
```

4.3.5 View 层

View 层包含两个 JSP 页面，一个是信息输入页面 userAdd.jsp，一个是信息显示页面 userList.jsp。

在 userAdd.jsp 页面中将 Map 类型的 hobbys 绑定到了 checkboxes 上，将 String[]类型

的 carrers 和 houseRegisters 绑定到 select 上，实现通过 option 标签对 select 添加选项，同时 method 方法需指定为 post 来避免中文乱码问题。

在 userList.jsp 页面中使用 JSTL 标签遍历集合中的用户信息，JSTL 相关知识参见本书的相关内容。

userAdd.jsp 的代码如下：

```
<%@ page language="java" contentType="text/html; charset=UTF-8"
pageEncoding="UTF-8"%>
<%@ taglib prefix="form" uri="http://www.springframework.org/tags/form"
%>
<%
String path = request.getContextPath();
String basePath = request.getScheme()+"://"+request.getServerName()+":"+
request.getServerPort()+path+"/";
%>
<!DOCTYPE html PUBLIC "-//W3C//DTD HTML 4.01 Transitional//EN"
"http://www.w3.org/TR/html4/loose.dtd">
<html>
<head>
<base href="<%=basePath%>">
<meta http-equiv="Content-Type" content="text/html; charset=UTF-8">
<title>Insert title here</title>
</head>
<body>
<form:form modelAttribute="user" method="post" action="user/save">
    <fieldset>
        <legend>添加一个用户</legend>
        <p>
            <label>用户名:</label>
            <form:input path="userName"/>
        </p>
        <p>
            <label>爱好:</label>
            <form:checkboxes items="${hobbys}" path="hobby" />
        </p>
        <p>
            <label>朋友:</label>
            <form:checkbox path="friends" value="张三"/>张三
            <form:checkbox path="friends" value="李四"/>李四
            <form:checkbox path="friends" value="王五"/>王五
            <form:checkbox path="friends" value="赵六"/>赵六
        </p>
        <p>
            <label>职业:</label>
            <form:select path="carrer">
```

```
                <option/>请选择职业
                <form:options items="${carrers }"/>
            </form:select>
        </p>
        <p>
            <label>户籍:</label>
            <form:select path="houseRegister">
                <option/>请选择户籍
                <form:options items="${houseRegisters }"/>
            </form:select>
        </p>
        <p>
            <label>个人描述:</label>
            <form:textarea path="remark" rows="5"/>
        </p>
        <p id="buttons">
            <input id="reset" type="reset">
            <input id="submit" type="submit" value="添加">
        </p>
    </fieldset>
</form:form>
</body>
</html>
```

userList.jsp 的代码如下：

```
<%@ page language="java" contentType="text/html; charset=UTF-8"
pageEncoding="UTF-8"%>
<%@ taglib uri="http://java.sun.com/jsp/jstl/core" prefix="c" %>
<%
String path = request.getContextPath();
String basePath = request.getScheme()+"://"+request.getServerName()+":"+
request.getServerPort()+path+"/";
%>
<!DOCTYPE html PUBLIC "-//W3C//DTD HTML 4.01 Transitional//EN"
"http://www.w3.org/TR/html4/loose.dtd">
<html>
<head>
<base href="<%=basePath%>">
<meta http-equiv="Content-Type" content="text/html; charset=UTF-8">
<title>Insert title here</title>
</head>
<body>
    <h1>用户列表</h1>
    <a href="<c:url value="user/input"/>">继续添加</a>
    <table>
```

```
        <tr>
            <th>用户名</th>
            <th>兴趣爱好</th>
            <th>朋友</th>
            <th>职业</th>
            <th>户籍</th>
            <th>个人描述</th>
        </tr>
        <!-- JSTL 标签，请参考本书的相关内容 -->
        <c:forEach items="${users}" var="user">
            <tr>
                <td>${user.userName }</td>
                <td>
                   <c:forEach items="${user.hobby }" var="hobby">
                        ${hobby } 
                   </c:forEach>
                </td>
                <td>
                   <c:forEach items="${user.friends }" var="friend">
                        ${friend } 
                   </c:forEach>
                </td>
                <td>${user.carrer }</td>
                <td>${user.houseRegister }</td>
                <td>${user.remark }</td>
            </tr>
        </c:forEach>
    </table>
</body>
</html>
```

4.3.6 测试应用

通过地址 http://localhost:8080/ch4/user/input 测试应用，添加用户信息页面效果如图 4.2 所示。

如果在图 4.2 中，职业选择“IT 民工”时，添加失败。失败后还回到添加页面，输入过的信息不再输入，自动回填（必须结合 form 标签）。自动回填是数据绑定的一个优点。失败页面如图 4.3 所示。

在图 4.2 中输入正确信息，添加成功后，重定向到信息显示页面，效果如图 4.4 所示。

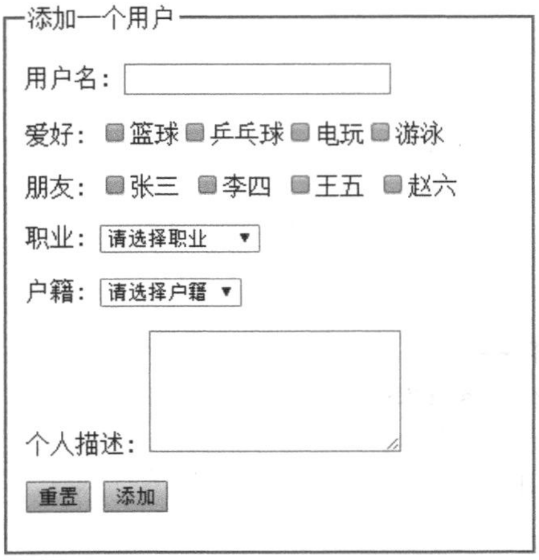

图 4.2　添加用户信息页面

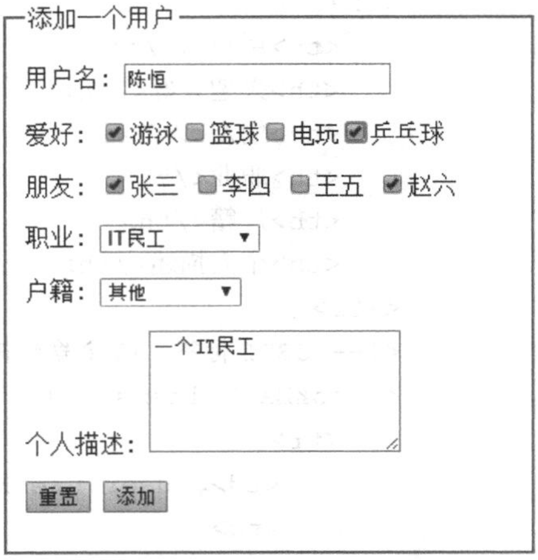

图 4.3　添加用户信息失败页面

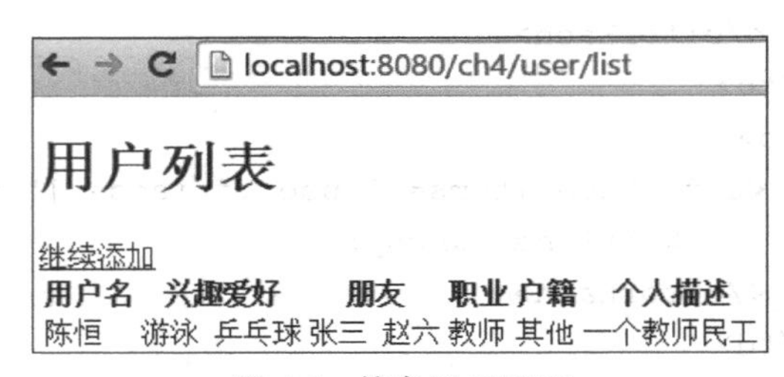

图 4.4　信息显示页面

4.4 实践环节

参考应用 ch4 创建应用 practice44。在应用 practice44 中创建两个视图页面 addGoods.jsp 和 goodsList.jsp。addGoods.jsp 页面的显示效果如图 4.5 所示，goodsList.jsp 页面的显示效果如图 4.6 所示。

图 4.5　添加商品页面

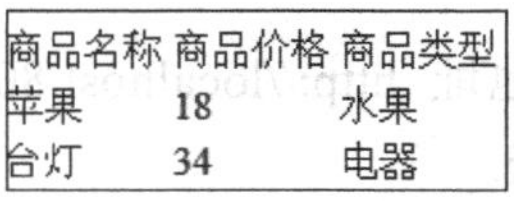

商品名称	商品价格	商品类型
苹果	18	水果
台灯	34	电器

图 4.6　商品显示页面

具体要求：

（1）商品类型由控制器类 GoodsController 的方法 inputGoods 进行初始化。GoodsController 类中共有三个方法：inputGoods、addGoods 和 listGoods。

（2）使用 Goods 模型类封装请求参数。

（3）使用 Service 层，在 Service 的实现类中，使用静态集合变量模拟数据库存储商品信息，在控制器类中使用@Autowired 注解 Service。

（4）通过地址 http://localhost:8080/practice44/goods/input 访问 addGoods.jsp 页面。

（5）其他的注意事项参见应用 ch4。

4.5 本 章 小 结

本章介绍了 Spring MVC 的数据绑定和表单标签，包括数据绑定的原理以及如何使用表单标签。最后给出了一个数据绑定应用示例，大致演示了数据绑定在实际开发中的使用。

习 题 4

1．举例说明数据绑定的优点。

2．Spring MVC 有哪些表单标签？其中，可以绑定集合数据的标签有哪些？

第5章

数 据 验 证

学习目的与要求

本章重点讲解 Spring MVC 框架的输入验证体系。通过本章的学习，理解输入验证的流程，能够利用 Spring 的自带验证框架和 JSR 303（Java 验证规范）对数据进行验证。

本章主要内容

- 数据验证概述
- Spring 验证
- JSR 303 验证

所有用户的输入一般都是随意的，为了保证数据的合法性，数据验证是所有 Web 应用必须处理的问题。在 Spring MVC 框架中，有两种方法可以验证输入数据：一是利用 Spring 自带的验证框架，一是利用 JSR 303 实现。

5.1 数据验证概述

数据验证分为客户端验证和服务器端验证，客户端验证主要是过滤正常用户的误操作，主要通过 JavaScript 代码完成；服务器端验证是整个应用阻止非法数据的最后防线，主要通过在应用中编程实现。

5.1.1 客户端验证

大多数情况下，使用 JavaScript 进行客户端验证的步骤如下：

（1）编写验证函数；

（2）在提交表单的事件中调用验证函数；

（3）根据验证函数来判断是否进行表单提交。

客户端验证可以过滤用户的误操作，是第一道防线，一般使用 JavaScript 代码实现。仅有客户端验证还是不够的。攻击者还可以绕过客户端验证直接进行非法输入，这样可能会引起系统的异常，为了确保数据的合法性，防止用户通过非正常手段提交错误信息。

所以必须加上服务器端的验证。

5.1.2 服务器端验证

Spring MVC 的 Converter 和 Formatter 在进行类型转换时，是将输入数据转换成领域对象的属性值（一种 Java 类型），一旦成功，服务器端验证器就会介入。也就是说，在 Spring MVC 框架中，先进行数据类型转换，再进行服务器端验证。

服务器端验证对于系统的安全性、完整性、健壮性起到了至关重要的作用。在 Spring MVC 框架中，可以利用 Spring 自带的验证框架验证数据，也可以利用 JSR 303 实现数据验证。

5.2 Spring 验证器

5.2.1 Validator 接口

创建自定义 Spring 验证器，需要实现 org.springframework.validation.Validator 接口。该接口有两个接口方法：

```
boolean supports(Class<?> klass)
void validate(Object object, Errors errors)
```

supports 方法返回 True 时，验证器可以处理指定的 Class。validate 方法的功能是验证目标对象 object，并将验证错误消息存入 errors 对象。

往 errors 对象存入错误消息的方法是 reject 或 rejectValue 方法。这两个方法的部分重载方法如下：

```
void reject(String errorCode)
void reject(String errorCode, String defaultMessage)
void rejectValue(String field, String errorCode)
void rejectValue(String field, String errorCode, String defaultMessage)
```

一般情况下，只需要给 reject 或 rejectValue 方法一个错误代码，Spring MVC 框架就会在消息属性文件中查找错误代码，获取相应错误消息。具体示例如下：

```
if(goods.getGprice() > 100 || goods.getGprice() < 0){
      errors.rejectValue("gprice", "gprice.invalid");//gprice.invalid
                                                     //为错误代码
}
```

5.2.2 ValidationUtils 类

org.springframework.validation.ValidationUtils 是一个工具类，该类中有几个方法可以

帮助判定值是否为空。

例如：

```
if(goods.getGname() == null || goods.getGname().isEmpty()){
        errors.rejectValue("gname", "goods.gname.required")
}
```

上述 if 语句可以使用 ValidationUtils 类的 rejectIfEmpty 方法，代码如下：

```
// errors 为 Errors 对象
// gname 为 goods 对象的属性
ValidationUtils.rejectIfEmpty(errors, "gname", "goods.gname.required");
```

再如：

```
if(goods.getGname() == null || goods.getGname().trim().isEmpty()){
        errors.rejectValue("gname", "goods.gname.required")
}
```

上述if语句可以编写成：

```
//gname 为 goods 对象的属性
ValidationUtils.rejectIfEmptyOrWhitespace(errors, "gname", "goods.
                                                         gname.required");
```

5.2.3 验证示例

本节使用一个应用 ch5a 讲解 Spring 验证器的编写及使用。ch5a 的项目结构如图 5.1 所示。

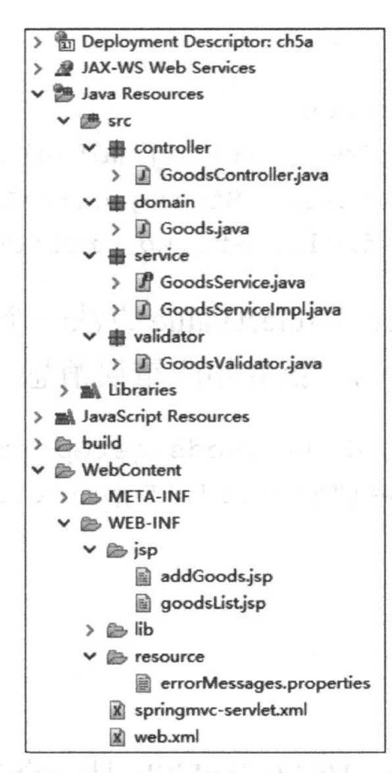

图 5.1 ch5a 的项目结构

该应用中有个数据输入页面 addGoods.jsp，效果如图 5.2 所示；有个数据显示页面 goodsList.jsp，效果如图 5.3 所示。

图 5.2 数据输入页面

图 5.3 数据显示页面

编写一个实现 org.springframework.validation.Validator 接口的验证器类 GoodsValidator，验证要求如下：

（1）商品名和商品详情不能为空。

（2）商品价格在 0～100 之间。

（3）创建日期不能在系统日期之后。

根据上述要求，按照如下步骤完成应用 ch5a。

1. 编写模型类

定义领域模型类 Goods，封装输入参数。在该类中使用@DateTimeFormat(pattern= "yyyy-MM-dd")格式化创建日期。模型类 Goods 的具体代码如下：

```
package domain;
import java.util.Date;
import org.springframework.format.annotation.DateTimeFormat;
public class Goods {
    private String gname;
    private String gdescription;
    private double gprice;
    //日期格式化（需要在配置文件配置 FormattingConversionServiceFactoryBean）
    @DateTimeFormat(pattern="yyyy-MM-dd")
    private Date gdate;
    public String getGname() {
        return gname;
    }
    public void setGname(String gname) {
        this.gname = gname;
    }
    public String getGdescription() {
        return gdescription;
    }
```

```
    public void setGdescription(String gdescription) {
        this.gdescription = gdescription;
    }
    public double getGprice() {
        return gprice;
    }
    public void setGprice(double gprice) {
        this.gprice = gprice;
    }
    public Date getGdate() {
        return gdate;
    }
    public void setGdate(Date gdate) {
        this.gdate = gdate;
    }
}
```

2. 编写验证器类

编写实现 org.springframework.validation.Validator 接口的验证器类 GoodsValidator，使用@Component 注解将 GoodsValidator 类声明为组件。具体代码如下：

```
package validator;
import java.util.Date;
import org.springframework.stereotype.Component;
import org.springframework.validation.Errors;
import org.springframework.validation.ValidationUtils;
import org.springframework.validation.Validator;
import domain.Goods;
@Component
public class GoodsValidator implements Validator{
    @Override
    public boolean supports(Class<?> klass) {
        //要验证的Model，返回值为False则不验证
       return Goods.class.isAssignableFrom(klass);
    }
    @Override
    public void validate(Object object, Errors errors) {
        Goods goods = (Goods)object;//object要验证的对象
        //goods.gname.required是错误消息属性文件中的编码（国际化后，对应的是国
际化的信息）
        ValidationUtils.rejectIfEmpty(errors, "gname",
"goods.gname.required");
        ValidationUtils.rejectIfEmpty(errors, "gdescription",
"goods.gdescription.required");
        if(goods.getGprice() > 100 || goods.getGprice() < 0){
```

```
            errors.rejectValue("gprice", "gprice.invalid");
        }
        Date goodsDate = goods.getGdate();
        //在系统时间之后
        if(goodsDate != null && goodsDate.after(new Date())){
            errors.rejectValue("gdate", "gdate.invalid");
        }
    }
}
```

3. 编写错误消息属性文件

在/WEB-INF/resource 目录下，编写属性文件 errorMessages.properties。文件内容如下：

```
goods.gname.required=请输入商品名称。
goods.gdescription.required=请输入商品详情。
gprice.invalid=价格在 0~100 之间。
gdate.invalid=创建日期不能在系统日期之后。
```

Unicode 编码（Eclipse 带有将汉字转换成 Unicode 编码的功能）的属性文件内容如下：

```
goods.gname.required=\u8BF7\u8F93\u5165\u5546\u54C1\u540D\u79F0\u3002
goods.gdescription.required=\u8BF7\u8F93\u5165\u5546\u54C1\u8BE6\u60C5
\u3002
gprice.invalid=\u4EF7\u683C\u57280-100\u4E4B\u95F4\u3002
gdate.invalid=\u521B\u5EFA\u65E5\u671F\u4E0D\u80FD\u5728\u7CFB\u7EDF\u
65E5\u671F\u4E4B\u540E\u3002
```

属性文件创建完成后，想要告诉 Spring MVC 从该文件中获取错误消息，则需要在配置文件中声明一个 messageSource bean，具体代码如下：

```
<!-- 配置消息属性文件 -->
<bean id="messageSource"
      class="org.springframework.context.support.ReloadableResource_
BundleMessageSource">
      <property name="basename" value="/WEB-INF/resource/error Messages"/>
</bean>
```

4. 编写 Service 层

在 Service 层中编写一个 GoodsService 接口和 GoodsServiceImpl 实现类。具体代码如下：

```
package service;
import java.util.ArrayList;
import domain.Goods;
public interface GoodsService {
```

```
    boolean save(Goods g);
    ArrayList<Goods> getGoods();
}
package service;
import java.util.ArrayList;
import org.springframework.stereotype.Service;
import domain.Goods;
@Service
public class GoodsServiceImpl implements GoodsService{
    //使用静态集合变量users模拟数据库
    private static ArrayList<Goods> goods = new ArrayList<Goods>();
    @Override
    public boolean save(Goods g) {
        goods.add(g);
        return true;
    }
    @Override
    public ArrayList<Goods> getGoods() {
        return goods;
    }
}
```

5. 编写控制器类

编写控制器类 GoodsController，在该类中使用@Resource 注解注入自定义验证器。另外，控制器类中包含两个处理请求的方法，具体代码如下：

```
package controller;
import javax.annotation.Resource;
import org.apache.commons.logging.Log;
import org.apache.commons.logging.LogFactory;
import org.springframework.beans.factory.annotation.Autowired;
import org.springframework.stereotype.Controller;
import org.springframework.ui.Model;
import org.springframework.validation.BindingResult;
import org.springframework.validation.Validator;
import org.springframework.web.bind.annotation.ModelAttribute;
import org.springframework.web.bind.annotation.RequestMapping;
import domain.Goods;
import service.GoodsService;
@Controller
@RequestMapping("/goods")
public class GoodsController {
    //得到一个用来记录日志的对象，这样打印信息的时候能够标记打印的是哪个类的信息
    private static final Log logger = LogFactory.getLog(GoodsController.
class);
```

```
    @Autowired
    private GoodsService goodsService;
    //注解验证器相当于 GoodsValidator validator = new  GoodsValidator();
    @Resource
    private Validator validator;
    @RequestMapping("/input")
    public String input(Model model){
        // 如果model中没有goods属性,addGoods.jsp会抛出异常,
        // 因为表单标签无法找到modelAttribute属性指定的form
        // backing object
        model.addAttribute("goods", new Goods());
        return "addGoods";
    }
    @RequestMapping("/save")
    public String save(@ModelAttribute Goods goods, BindingResult result,
Model model){
        this.validator.validate(goods, result);//添加验证
        if (result.hasErrors()) {
            return "addGoods";
        }
        goodsService.save(goods);
        logger.info("添加成功");
        model.addAttribute("goodsList", goodsService.getGoods());
        return "goodsList";
    }
}
```

6. 编写配置文件

配置文件 springmvc-servlet.xml 的代码如下：

```
<?xml version="1.0" encoding="UTF-8"?>
<beans xmlns="http://www.springframework.org/schema/beans"
   xmlns:xsi="http://www.w3.org/2001/XMLSchema-instance"
   xmlns:p="http://www.springframework.org/schema/p"
   xmlns:context="http://www.springframework.org/schema/context"
   xmlns:mvc="http://www.springframework.org/schema/mvc"
   xsi:schemaLocation="
    http://www.springframework.org/schema/beans
    http://www.springframework.org/schema/beans/spring-beans.xsd
      http://www.springframework.org/schema/context
      http://www.springframework.org/schema/context/spring-context.xsd
      http://www.springframework.org/schema/mvc
      http://www.springframework.org/schema/mvc/spring-mvc.xsd">
   <!-- 使用扫描机制，扫描包 -->
   <context:component-scan base-package="controller"/>
   <context:component-scan base-package="service"/>
```

```
    <context:component-scan base-package="validator"/>
    <!-- 注册格式化转换器 -->
    <bean id="conversionService" class="org.springframework.format.
     support.FormattingConversionServiceFactoryBean">
     <property name="formatters">
         <set>
             <!-- 注册自定义格式化转换器 -->
         </set>
     </property>
    </bean>
    <mvc:annotation-driven conversion-service="conversionService"/>
    <!-- 配置消息属性文件 -->
    <bean id="messageSource" class="org.springframework.context.support.
ReloadableResourceBundleMessageSource">
     <property name="basename" value="/WEB-INF/resource/errorMessages"/>
    </bean>
     <!-- 配置视图解析器 -->
     <bean class="org.springframework.web.servlet.view.InternalResource_
ViewResolver"
            id="internalResourceViewResolver">
       <!-- 前缀 -->
       <property name="prefix" value="/WEB-INF/jsp/" />
       <!-- 后缀 -->
       <property name="suffix" value=".jsp" />
  </bean>
</beans>
```

7. 编写视图 JSP 文件

应用 ch5a 中有两个视图文件：addGoods.jsp 和 goodsList.jsp。
addGoods.jsp 的代码如下：

```
<%@   page   language="java"   contentType="text/html;   charset=UTF-8"
pageEncoding="UTF-8"%>
<%@ taglib prefix="form" uri="http://www.springframework.org/tags/form" %>
<%
String path = request.getContextPath();
String basePath = request.getScheme()+"://"+request.getServerName()+":"+
request.getServerPort()+path+"/";
%>
<!DOCTYPE   html   PUBLIC   "-//W3C//DTD   HTML   4.01   Transitional//EN"
"http://www.w3.org/TR/html4/loose.dtd">
<html>
<head>
<base href="<%=basePath%>">
<meta http-equiv="Content-Type" content="text/html; charset=UTF-8">
```

```
<title>Insert title here</title>
</head>
<body>
    <form:form modelAttribute="goods" action="goods/save" method="post">
        <fieldset>
      <legend>添加一件商品</legend>
      <p>
          <label>商品名:</label>
          <form:input path="gname"/>
      </p>
      <p>
          <label>商品详情:</label>
          <form:input path="gdescription"/>
      </p>
      <p>
          <label>商品价格:</label>
          <form:input path="gprice"/>
      </p>
      <p>
          <label>创建日期:</label>
          <form:input path="gdate"/>(yyyy-MM-dd)
      </p>
      <p id="buttons">
          <input id="reset" type="reset">
          <input id="submit" type="submit" value="添加">
      </p>
    </fieldset>
    <!-- 取出所有验证错误 -->
    <form:errors path="*"/>
    </form:form>
</body>
</html>
```

goodsList.jsp 的代码如下：

```
<%@ page language="java" contentType="text/html; charset=UTF-8"
pageEncoding="UTF-8"%>
<%@ taglib uri="http://java.sun.com/jsp/jstl/core" prefix="c" %>
<%
String path = request.getContextPath();
String basePath = request.getScheme()+"://"+request.getServerName()+":"+
request.getServerPort()+path+"/";
%>
<!DOCTYPE html PUBLIC "-//W3C//DTD HTML 4.01 Transitional//EN"
"http://www.w3.org/TR/html4/loose.dtd">
<html>
```

```
<head>
<base href="<%=basePath%>">
<meta http-equiv="Content-Type" content="text/html; charset=UTF-8">
<title>Insert title here</title>
</head>
<body>
    <table>
        <tr>
            <td>商品名</td>
            <td>商品详情</td>
            <td>商品价格</td>
            <td>创建日期</td>
        </tr>
        <c:forEach items="${goodsList }" var="goods">
            <tr>
                <td>${goods.gname }</td>
                <td>${goods.gdescription }</td>
                <td>${goods.gprice }</td>
                <td>${goods.gdate }</td>
            </tr>
        </c:forEach>
    </table>
</body>
</html>
```

8. 测试应用

通过地址 http://localhost:8080/ch5a/goods/input 测试应用。

5.2.4 实践环节

参考第 5.2.3 节，创建应用 practice524，应用中有个输入页面，效果如图 5.4 所示。在该应用中创建一个实现 org.springframework.validation.Validator 接口的验证器类，验证器对输入页面的输入项进行验证。要求如下：

（1）所有输入项不能为空。

（2）年龄必须在 18 至 65 岁之间。

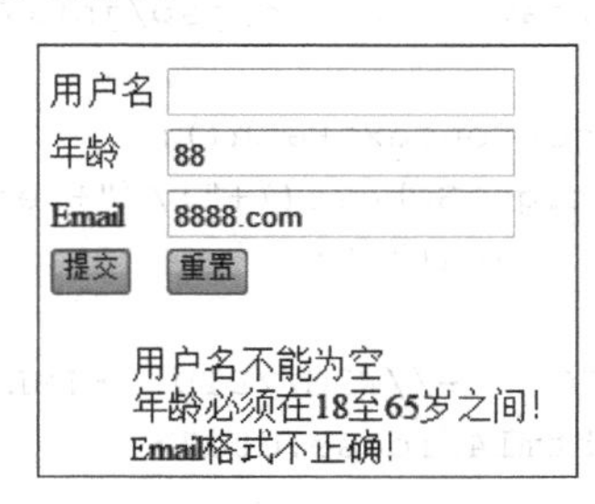

图 5.4 实践环节的输入页面

（3）Email 满足正常的格式。

5.3 JSR 303 验证

对于JSR 303验证，目前有两个实现，一个是Hibernate Validator，一个是Apache BVal。本书采用的是 Hibernate Validator，注意它和 Hibernate 无关，只是使用它进行数据验证。

5.3.1 JSR 303 验证配置

1．下载与安装 Hibernate Validator

可以通过地址 https://sourceforge.net/projects/hibernate/files/hibernate-validator/ 下载 Hibernate Validator，本书选择的是 hibernate-validator-5.4.0.Final-dist.zip。

首先，将下载的压缩包解压；然后，将\hibernate-validator-5.4.0.Final\dist 目录下的 hibernate-validator-5.4.0.Final.jar 和\hibernate-validator-5.4.0.Final\dist\lib\required 目录下的 classmate-1.3.1.jar、javax.el-3.0.1-b08.jar、jboss-logging-3.3.0.Final.jar、validation-api-1.1.0.Final.jar 复制到应用的\WEB-INF\lib 目录下。

2．配置属性文件与验证器

如果将验证错误消息放在属性文件中，那么需要在配置文件中配置属性文件，并将属性文件与 Hibernate Validator 关联，具体配置代码如下：

```
<!-- 配置消息属性文件 -->
<bean id="messageSource"
    class="org.springframework.context.support.ReloadableResource_
    BundleMessageSource">
    <!-- 资源文件名-->
    <property name="basenames">
       <list>
          <value>/WEB-INF/resource/errorMessages</value>
       </list>
       </property>
       <!-- 资源文件编码格式 -->
       <property name="fileEncodings" value="utf-8" />
       <!-- 对资源文件内容缓存时间，单位秒 -->
       <property name="cacheSeconds" value="120" />
   </bean>
   <!-- 注册校验器 -->
   <bean id="validator"
       class="org.springframework.validation.beanvalidation.Local_
       ValidatorFactoryBean">
       <!-- hibernate 校验器-->
```

```
        <property name="providerClass" value="org.hibernate.validator.
HibernateValidator" />
        <!-- 指定校验使用的资源文件，在文件中配置校验错误信息，如果不指定则默认使用
classpath 下的 ValidationMessages.properties -->
        <property name="validationMessageSource" ref="messageSource" />
    </bean>
    <!-- 开启 spring 的 Valid 功能 -->
    <mvc:annotation-driven conversion-service="conversionService" validator
="validator"/>
```

5.3.2 标注类型

JSR 303 不需要编写验证器，但需要利用它的标注类型在领域模型的属性上嵌入约束。

1．空检查

@Null：验证对象是否为 null。

@NotNull：验证对象是否不为 null，无法检查长度为 0 的字符串。

@NotBlank：检查约束字符串是不是 null，还有被 trim 后的长度是否大于 0，只针对字符串，且会去掉前后空格。

@NotEmpty：检查约束元素是否为 null 或者是 empty。

示例如下：

```
@NotBlank(message="{goods.gname.required}")//goods.gname.required 为属
性文件的错误代码
private String gname;
```

2．boolean 检查

@AssertTrue：验证 boolean 属性是否为 True。

@AssertFalse：验证 boolean 属性是否为 False。

示例如下：

```
@AssertTrue
private boolean isLogin;
```

3．长度检查

@Size(min=, max=)：验证对象(Array,Collection,Map,String)长度是否在给定的范围之内。

@Length(min=, max=)：验证字符串长度是否在给定的范围之内。

示例如下：

```
@Length(min=1,max=100)
```

```
private String gdescription;
```

4．日期检查

@Past：验证 Date 和 Calendar 对象是否在当前时间之前。
@Future：验证 Date 和 Calendar 对象是否在当前时间之后。
@Pattern：验证 String 对象是否符合正则表达式的规则。
示例如下：

```
@Past(message="{gdate.invalid}")
private Date gdate;
```

5．数值检查

@Min：验证 Number 和 String 对象是否大于等于指定的值。
@Max：验证 Number 和 String 对象是否小于等于指定的值。
@DecimalMax：被标注的值必须不大于约束中指定的最大值，这个约束的参数是一个通过 BigDecimal 定义的最大值的字符串表示，小数存在精度。
@DecimalMin：被标注的值必须不小于约束中指定的最小值，这个约束的参数是一个通过 BigDecimal 定义的最小值的字符串表示，小数存在精度。
@Digits：验证 Number 和 String 的构成是否合法。
@Digits(integer=,fraction=)：验证字符串是否符合指定格式的数字，integer 指定整数精度，fraction 指定小数精度。
@Range(min=, max=)：检查数字是否介于 min 和 max 之间。
@Valid：对关联对象进行校验，如果关联对象是个集合或者数组，那么对其中的元素进行校验，如果是一个 map，则对其中的值部分进行校验。
@CreditCardNumber：信用卡验证。
@Email：验证是否是邮件地址，如果为 null，不进行验证，通过验证。
示例如下：

```
@Range(min=0,max=100,message="{gprice.invalid}")
private double gprice;
```

5.3.3 验证示例

创建应用 ch5b，该应用实现的功能与第 5.2.3 节 ch5a 应用相同。ch5b 的项目结构如图 5.5 所示。

在应用 ch5b 中不需要创建验证器类 GoodsValidator。另外，Service 层、View 层以及错误消息属性文件都与 ch5a 应用的相同。

1．模型类

在模型类 Goods 中，利用 JSR 303 的标注类型对属性进行验证，具体代码如下：

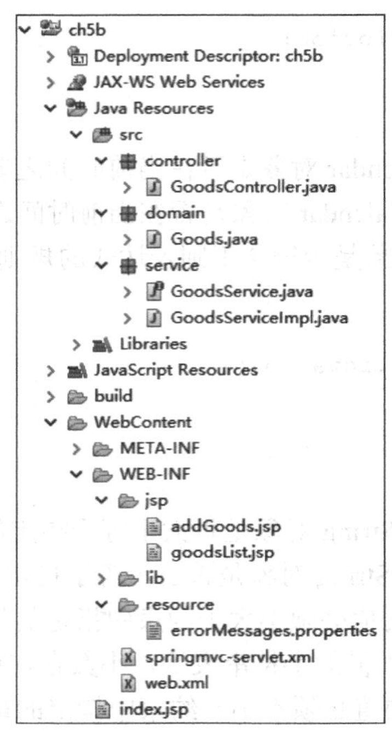

图 5.5 ch5b 的项目结构

```
package domain;
import java.util.Date;
import javax.validation.constraints.Past;
import org.hibernate.validator.constraints.NotBlank;
import org.hibernate.validator.constraints.Range;
import org.springframework.format.annotation.DateTimeFormat;
public class Goods {
    //goods.gname.required 错误消息 key（国际化后，对应的就是国际化的信息）
    @NotBlank(message="{goods.gname.required}")
    private String gname;
    @NotBlank(message="{goods.gdescription.required}")
    private String gdescription;
    @Range(min=0,max=100,message="{gprice.invalid}")
    private double gprice;
    //日期格式化（需要在配置文件配置 FormattingConversionServiceFactoryBean)
    @DateTimeFormat(pattern="yyyy-MM-dd")
    @Past(message="{gdate.invalid}")
    private Date gdate;
    public String getGname() {
        return gname;
    }
    public void setGname(String gname) {
        this.gname = gname;
    }
    public String getGdescription() {
```

```
        return gdescription;
    }
    public void setGdescription(String gdescription) {
        this.gdescription = gdescription;
    }
    public double getGprice() {
        return gprice;
    }
    public void setGprice(double gprice) {
        this.gprice = gprice;
    }
    public Date getGdate() {
        return gdate;
    }
    public void setGdate(Date gdate) {
        this.gdate = gdate;
    }
}
```

2. 控制器类

在控制器类 GoodsController 中，使用@Valid 对模型对象进行验证，具体代码如下：

```
package controller;
import javax.validation.Valid;
import org.apache.commons.logging.Log;
import org.apache.commons.logging.LogFactory;
import org.springframework.beans.factory.annotation.Autowired;
import org.springframework.stereotype.Controller;
import org.springframework.ui.Model;
import org.springframework.validation.BindingResult;
import org.springframework.web.bind.annotation.ModelAttribute;
import org.springframework.web.bind.annotation.RequestMapping;
import domain.Goods;
import service.GoodsService;
@Controller
@RequestMapping("/goods")
public class GoodsController {
    //得到一个用来记录日志的对象，这样打印信息的时候能够标记打印的是哪个类的信息
    private static final Log logger = LogFactory.getLog(GoodsController.
class);
    @Autowired
    private GoodsService goodsService;
    @RequestMapping("/input")
    public String input(Model model){
        //如果model中没有goods属性，addGoods.jsp会抛出异常，因为表单标签无法找到
```

```
        //modelAttribute 属性指定的 form backing object
        model.addAttribute("goods", new Goods());
        return "addGoods";
    }
    @RequestMapping("/save")
    public String save(@Valid @ModelAttribute Goods goods, BindingResult
result, Model model){
        if(result.hasErrors()){
            return "addGoods";
        }
        goodsService.save(goods);
        logger.info("添加成功");
        model.addAttribute("goodsList", goodsService.getGoods());
        return "goodsList";
    }
}
```

3. 配置文件

配置文件 springmvc-servlet.xml 的代码如下：

```
<?xml version="1.0" encoding="UTF-8"?>
<beans xmlns="http://www.springframework.org/schema/beans"
    xmlns:xsi="http://www.w3.org/2001/XMLSchema-instance"
    xmlns:p="http://www.springframework.org/schema/p"
    xmlns:context="http://www.springframework.org/schema/context"
    xmlns:mvc="http://www.springframework.org/schema/mvc"
    xsi:schemaLocation="
     http://www.springframework.org/schema/beans
     http://www.springframework.org/schema/beans/spring-beans.xsd
        http://www.springframework.org/schema/context
        http://www.springframework.org/schema/context/spring-context.xsd
        http://www.springframework.org/schema/mvc
        http://www.springframework.org/schema/mvc/spring-mvc.xsd">
    <!-- 使用扫描机制，扫描包 -->
    <context:component-scan base-package="controller"/>
    <context:component-scan base-package="service"/>
     <!-- 配置消息属性文件 -->
    <bean id="messageSource"
        class="org.springframework.context.support.ReloadableResource_
BundleMessageSource">
        <!-- 资源文件名-->
        <property name="basenames">
         <list>
            <value>/WEB-INF/resource/errorMessages</value>
         </list>
        </property>
```

```
        <!-- 资源文件编码格式 -->
        <property name="fileEncodings" value="utf-8" />
        <!-- 对资源文件内容缓存时间，单位秒 -->
        <property name="cacheSeconds" value="120" />
    </bean>
    <!-- 注册校验器 -->
    <bean id="validator"
        class="org.springframework.validation.beanvalidation.Local_
ValidatorFactoryBean">
        <!-- hibernate 校验器-->
        <property name="providerClass" value="org.hibernate.validator.
HibernateValidator" />
        <!-- 指定校验使用的资源文件，在文件中配置校验错误信息，如果不指定则默认使用
classpath 下的 ValidationMessages.properties -->
        <property name="validationMessageSource" ref="messageSource" />
    </bean>
    <!-- 开启 spring 的 Valid 功能 -->
    <mvc:annotation-driven conversion-service="conversionService" validator=
"validator"/>
     <!-- 注册格式化转换器 -->
    <bean id="conversionService" class="org.springframework.format._
support.FormattingConversionServiceFactoryBean">
     <property name="formatters">
         <set>
             <!-- 注册自定义格式化转换器 -->
         </set>
     </property>
    </bean>
     <!-- 配置视图解析器 -->
     <bean class="org.springframework.web.servlet.view.InternalResource_
          ViewResolver"id="internalResourceViewResolver">
       <!-- 前缀 -->
       <property name="prefix" value="/WEB-INF/jsp/" />
       <!-- 后缀 -->
       <property name="suffix" value=".jsp" />
  </bean>
</beans>
```

4．测试应用

通过地址 http://localhost:8080/ch5b/goods/input 测试应用 ch5b。

5.3.4 实践环节

参考第 5.3.3 节的应用 ch5b 创建应用 practice534，在该应用中使用 JSR 303 对输入

页面的输入项进行验证。要求如下：

（1）用户名不能为空，并且长度在 3～10。

（2）使用正则表达式验证手机号。

（3）生日满足日期格式，并且是在 1990-01-01—2015-07-31。

（4）输入页面的运行效果如图 5.6 所示。

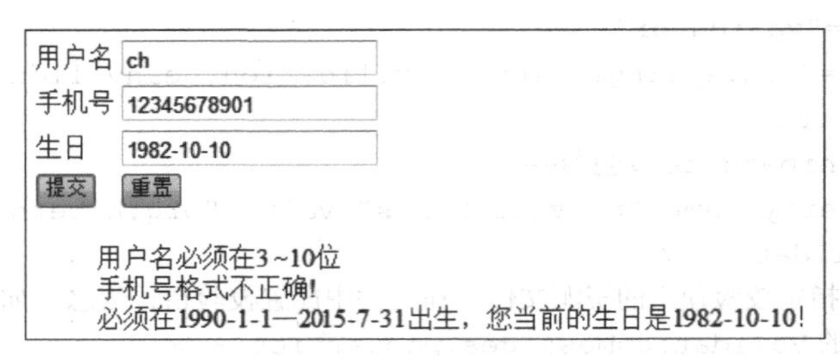

图 5.6 输入页面

5.4 本章小结

本章重点讲解了 Spring 验证的编写与 JSR 303 验证的使用方法。不管哪种验证方式，都需要注意验证流程。

习 题 5

1．如何创建 Spring 验证器类？

2．举例说明 JSR 303 验证的标注类型的使用方法。

第6章

国 际 化

学习目的与要求

本章重点讲解 Spring MVC 国际化的实现方法。通过本章的学习，读者应理解 Spring MVC 国际化的设计思想，掌握 Spring MVC 国际化的实现方法。

本章主要内容

- Java 国际化的思想
- Spring MVC 的国际化
- 用户自定义切换语言

国际化是商业软件系统的一个基本要求，因为当今的软件系统需要面对全球的浏览者。国际化的目的，就是根据用户的语言环境不同，输出与之相应的页面给用户，以示友好。

Spring MVC 的国际化主要有 JSP 页面信息国际化以及错误消息国际化。错误消息在第 5 章已讲解，本章主要介绍如何在 JSP 页面输出国际化消息。最后，本章将示范一个让用户自行选择语言的示例。

6.1 程序国际化概述

程序国际化已成为 Web 应用的基本要求。随着网络的发展，大部分的 Web 站点面对的已经不再是本地或者本国的浏览者，而是来自全世界各国各地区的浏览者，因此国际化成了 Web 应用不可或缺的一部分。

6.1.1 Java 国际化的思想

Java 国际化的思想是将程序中的信息放在资源文件中，程序根据支持的国家及语言环境，读取相应的资源文件。资源文件是 key-value 对，每个资源文件中的 key 是不变的，但 value 则随不同国家/语言变化。

在 Java 程序中的国际化主要通过两个类来完成。

- java.util.Locale：用于提供本地信息，通常称它为语言环境。不同的语言，不同的国家和地区采用不同的 Locale 对象来表示。
- java.util.ResourceBundle：该类称为资源包，包含了特定于语言环境的资源对象。当程序需要一个特定于语言环境的资源时（如字符串资源），程序可以从适合当前用户语言环境的资源包中加载它。采用这种方式，可以编写独立于用户语言环境的程序代码，而与特定语言环境相关的信息则通过资源包来提供。

为了实现 Java 程序的国际化，必须事先提供程序所需要的资源文件。资源文件的内容是由很多 key-value 对组成的，其中 key 是程序使用的部分，而 value 则是程序界面的显示。

资源文件的命名可以有如下三种形式：

- baseName.properties
- baseName_language.properties
- baseName_language_country.properties

baseName 是资源文件的基本名称，由用户自由定义。但是 language 和 country 就必须为 Java 所支持的语言和国家/地区代码。例如：

```
中国大陆： baseName_zh_CN.properties
美国： baseName_en_US.properties
```

Java 中的资源文件只支持 ISO-8859-1 编码格式字符，直接编写中文会出现乱码。可以使用 Java 命令 native2ascii.exe 解决资源文件的中文乱码问题。使用 Eclipse 编写资源属性文件，在保存资源文件时，Eclipse 自动执行 native2ascii.exe 命令。因此，在 Eclipse 中资源文件不会出现中文乱码问题。

6.1.2　Java 支持的语言和国家

java.util.Locale 类的常用构造方法如下：

- public Locale(String language)
- public Locale(String language,String country)

其中 language 表示语言，它的取值是由小写的两个字母组成的语言代码。country 表示国家或地区，它的取值是由大写的两个字母组成的国家或地区代码。

实际上，Java 并不能支持所有国家和语言，如果需要获取 Java 所支持的语言和国家，开发者可以通过调用 Locale 类的 getAvailableLocales 方法获取，该方法返回一个 Locale 数组，该数组里包含了 Java 所支持的语言和国家。

下面的 Java 程序简单示范了如何获取 Java 所支持的国家和语言：

```
import java.util.Locale;
public class Test {
    public static void main(String[] args) {
        // 返回 Java 所支持的语言和国家的数组
        Locale locales[] = Locale.getAvailableLocales();
```

```
            //遍历数组元素，依次获取所支持的国家和语言
            for (int i = 0; i < locales.length; i++) {
                //打印出所支持的国家和语言
                System.out.println(locales[i].getDisplayCountry() + "="
                        + locales[i].getCountry() + " "
                        + locales[i].getDisplayLanguage() + "="
                        + locales[i].getLanguage());
            }
        }
    }
```

程序运行结果如图 6.1 所示。

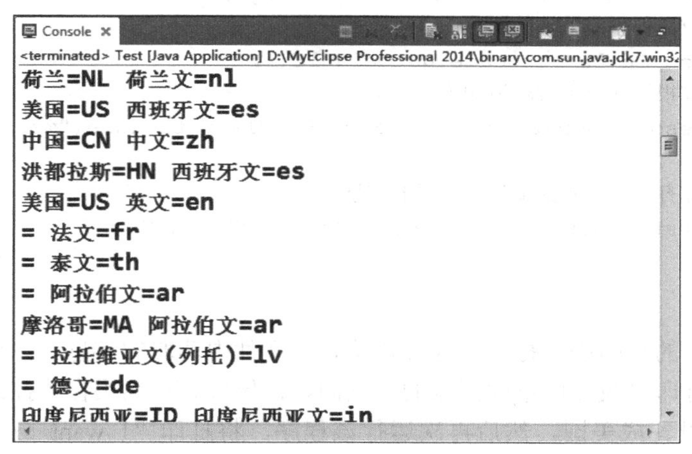

图 6.1 Java 所支持国家和语言

6.1.3 Java 程序国际化

假设有如下的简单 Java 程序：

```
public class TestI18N {
    public static void main(String[] args) {
        System.out.println("我要向不同国家的人民问好：您好！");
    }
}
```

为了让该程序支持国际化，需要将“我要向不同国家的人民问好：您好！”对应不同语言环境的字符串定义在不同的资源文件中。

在 Web 应用的 src 目录下新建文件 messageResource_zh_CN.properties 和 messageResource_en_US.properties。

其次，给资源文件 messageResource_zh_CN.properties 添加“我要向不同国家的人民问好：您好！”的内容，可看到如图 6.2 所示的效果。

```
hello=\u6211\u8981\u5411\u4E0D\u540C\u56
```

图 6.2　Unicode 编码资源文件

图 6.2 显示的内容看似是很多乱码，实际是 Unicode 编码文件内容。至此，资源文件 messageResource_zh_CN.properties 创建完成。同理，创建资源文件 messageResource_en_US.properties，具体步骤不再赘述。

现在将 TestI18N.java 程序修改成如下形式：

```
import java.util.Locale;
import java.util.ResourceBundle;
public class TestI18N {
    public static void main(String[] args) {
        //取得系统默认的国家语言环境
        Locale lc = Locale.getDefault();
        //根据国家语言环境加载资源文件
        ResourceBundle rb = ResourceBundle.getBundle("messageResource",
lc);
        //打印出从资源文件中取得的信息
        System.out.println(rb.getString("hello"));
    }
}
```

上面程序中的打印语句打印的内容是从资源文件中读取的信息。如果在中文环境下运行程序，将打印“我要向不同国家的人民问好：您好！”；如果在“控制面板”中将机器的语言环境设置成美国，然后再次运行该程序，将打印“I want to say hello to all world!”。

需要注意的是，如果程序找不到对应国家/语言的资源文件时，系统该怎么办？假设以简体中文环境为例，先搜索如下文件：

```
messageResource_zh_CN.properties
```

如果没有找到国家/语言都匹配的资源文件，再搜索语言匹配文件，即搜索如下文件：

```
messageResource_zh.properties
```

如果上面的文件还没有搜索到，则搜索 baseName 匹配的文件，即搜索如下文件：

```
messageResource.properties
```

如果上面三个文件都找不到，则系统将出现异常。

6.1.4　带占位符的国际化信息

在资源文件中的消息文本可以带有参数，例如：

```
welcome={0}，欢迎学习 Spring MVC。
```

花括号中的数字是一个占位符，可以被动态的数据替换。在消息文本中的占位符可以使用 0 到 9 的数字，也就是说，消息文本的参数最多可以有 10 个。例如：

```
welcome={0}，欢迎学习 Spring MVC，今天是星期{1}。
```

要替换消息文本中的占位符，可以使用 java.text.MessageFormat 类，该类提供了一个静态方法 format()，用来格式化带参数的文本，format() 方法定义如下：

```
public static String format(String pattern,Object …arguments)
```

其中，pattern 字符串就是一个带占位符的字符串，消息文本中的数字占位符将按照方法参数的顺序（从第二个参数开始）而被替换。

替换占位符的示例代码如下：

```
import java.text.MessageFormat;
import java.util.Locale;
import java.util.ResourceBundle;
public class TestFormat {
    public static void main(String[] args) {
        //取得系统默认的国家语言环境
        Locale lc = Locale.getDefault();
        //根据国家语言环境加载资源文件
        ResourceBundle rb = ResourceBundle.getBundle("messageResource", lc);
        //从资源文件中取得的信息
        String msg = rb.getString("welcome");
        //替换消息文本中的占位符，消息文本中的数字占位符将按照参数的顺序
        //(从第二个参数开始)而被替换，即"我"替换{0}，"5"替换{1}
        String msgFor = MessageFormat.format(msg, "我","5");
        System.out.println(msgFor);
    }
}
```

6.1.5 实践环节

编写一个 Java 应用程序，在该程序中从资源文件 messageResource_zh_CN.properties 中读取消息文本“practice615=今天{0}很高兴，{1}也不错，明天就是星期{2}了。”。对应的英文资源文件是 messageResource_en_US.properties，消息文本是“practice615= Today,{0} is very glad,{1} is too good,tomorrow will be {2}.”。

6.2 Spring MVC 的国际化

Spring MVC 的国际化是建立在 Java 国际化的基础之上的，Spring MVC 框架的底层国际化与 Java 国际化是一致的，作为一个良好的 MVC 框架，Spring MVC 将 Java 国际

化的功能进行了封装和简化，开发者使用起来会更加简单快捷。

由第 6.1 节可知，国际化和本地化应用程序时，需要具备以下两个条件：

（1）将文本信息放到资源属性文件中。

（2）选择和读取正确位置的资源属性文件。

下面讲解第（2）个条件的实现。

6.2.1 Spring MVC 加载资源属性文件

在 Spring MVC 中，不能直接使用 ResourceBundle 加载资源属性文件，而是利用 bean（messageSource）告知 Spring MVC 框架要将资源属性文件放到哪里。示例代码如下：

```
<bean id="messageSource"
    class="org.springframework.context.support.
ReloadableResourceBundleMessageSource">
    <!-- <property name="basename" value="classpath:messages" /> -->
    <property name="basename" value="/WEB-INF/resource/messages" />
</bean>
```

上述 bean 配置的是国际化资源文件的路径，“classpath:messages”指的是 classpath 路径下的 messages_zh_CN.properties 文件和 messages_en_US.properties 文件。当然也可以将国际化资源文件放在其他的路径下，如/WEB-INF/resource/messages。

另外，“messageSource”bean 是由 ReloadableResourceBundleMessageSource 类实现的，它是不能重新加载的，如果修改了国际化资源文件，需要重启 JVM。

最后，还需要注意如果有一组属性文件，则用 basenames 替换 basename，示例代码如下：

```
<bean id="messageSource"
    class="org.springframework.context.support.
ReloadableResourceBundleMessageSource">
    <property name="basenames">
        <list>
            <value>/WEB-INF/resource/messages</value>
            <value>/WEB-INF/resource/labels</value>
        </list>
    </property>
</bean>
```

6.2.2 语言区域的选择

在 Spring MVC 中，可以使用语言区域解析器 bean 选择语言区域。该 bean 由三个常见实现：AcceptHeaderLocaleResolver、SessionLocaleResolver 以及 CookieLocaleResolver。

- AcceptHeaderLocaleResolver

根据浏览器 Http Header 中的 accept-language 域判定（accept-language 域中一般包含

了当前操作系统的语言设定，可通过 HttpServletRequest.getLocale 方法获得此域的内容)。改变 Locale 是不支持的，即不能调用 LocaleResolver 接口的 setLocale(HttpServletRequest request, HttpServletResponse response, Locale locale)方法设置 Locale。

- SessionLocaleResolver

根据用户本次会话过程中的语言设定决定语言区域（如用户进入首页时选择语言种类，则此次会话周期内统一使用该语言设定)。

- CookieLocaleResolver

根据 Cookie 判定用户的语言设定（Cookie 中保存着用户前一次的语言设定参数)。

由上述分析可知，SessionLocaleResolver 实现，比较方便用户选择喜欢的语言种类，本章使用该方法进行国际化实现。

下面是使用 SessionLocaleResolver 实现的 bean 定义：

```
<bean id="localeResolver" class="org.springframework.web.servlet.i18n.
SessionLocaleResolver">
    <property name="defaultLocale" value="zh_CN"></property>
 </bean>
```

如果采用基于 SessionLocaleResolver 和 CookieLocaleResolver 国际化实现，必须配置 LocaleChangeInterceptor 拦截器，示例代码如下：

```
<mvc:interceptors>
   <bean class="org.springframework.web.servlet.i18n.LocaleChange_
Interceptor"/>
</mvc:interceptors>
```

6.2.3 使用 message 标签显示国际化信息

在 Spring MVC 框架中，可以使用 Spring 的 message 标签在 JSP 页面中显示国际化消息。使用 message 标签时，需要在 JSP 页面最前面使用 taglib 指令声明 spring 标签，代码如下：

```
<%@taglib prefix="spring" uri="http://www.springframework.org/tags" %>
```

message 标签有如下常用属性。

- code：获得国际化消息的 key。
- arguments：代表该标签的参数。如替换消息中的占位符，示例代码为

```
<spring:message code="third" arguments="888,999" />，third 对应的消息有两个占位符{0}和{1}。
```

- argumentSeparator：用来分隔该标签参数的字符，默认为逗号。
- text：如果 code 属性不存在，或指定的 key 无法获取消息时，所显示的默认文本信息。

6.3 用户自定义切换语言示例

在许多成熟的商业软件系统中，可以让用户自由切换语言，而不是修改浏览器的语言设置。一旦用户选择了自己需要使用的语言环境，整个系统的语言环境将一直是这种语言环境。Spring MVC 也可以允许用户自行选择程序语言。

创建 Spring MVC 应用 ch6，ch6 的项目结构如图 6.3 所示。

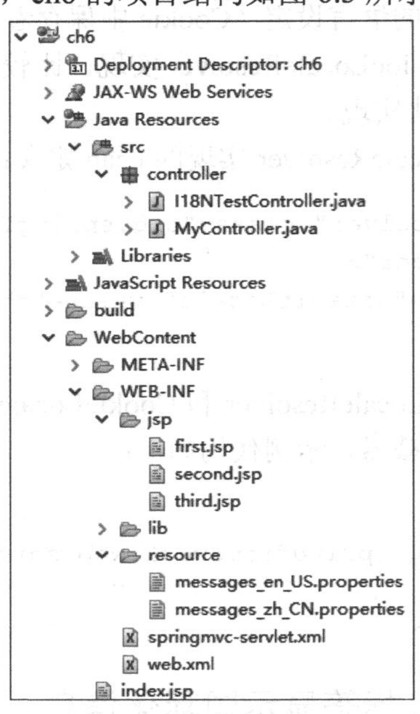

图 6.3 ch6 的项目结构

在该应用中使用 SessionLocaleResolver 实现国际化，具体步骤如下。

1. 编写国际化资源文件

在/WEB-INF/resource/目录下，编写中英文资源文件 messages_en_US.properties 和 messages_zh_CN.properties。

messages_en_US.properties 的内容如下：

```
first = first
second = second
third = {0} third {1}
language.en = English
language.cn = Chinese
```

messages_zh_CN.properties 的内容如下：

```
first = \u7B2C\u4E00\u9875
```

```
second = \u7B2C\u4E8C\u9875
third = {0} \u7B2C\u4E09\u9875 {1}
language.cn = \u4E2D\u6587
language.en = \u82F1\u6587
```

2. 编写视图 JSP 文件

应用中包含 4 个 JSP 文件：index.jsp、first.jsp、second.jsp 和 third.jsp。

index.jsp 的代码如下：

```
<% response.sendRedirect("my/first"); %>
```

first.jsp 的代码如下：

```
<%@ page language="java" contentType="text/html; charset=UTF-8"
pageEncoding="UTF-8"%>
<%@taglib prefix="spring" uri="http://www.springframework.org/tags" %>
<%
String path = request.getContextPath();
String basePath = request.getScheme()+"://"+request.getServerName()+":"+
request.getServerPort()+path+"/";
%>
<!DOCTYPE html PUBLIC "-//W3C//DTD HTML 4.01 Transitional//EN"
"http://www.w3.org/TR/html4/loose.dtd">
<html>
<head>
<base href="<%=basePath%>">
<meta http-equiv="Content-Type" content="text/html; charset=UTF-8">
<title><spring:message code="first"/></title>
</head>
<body>
    <a href="i18nTest?locale=zh_CN"><spring:message code="language.cn"
/></a> --
   <a href="i18nTest?locale=en_US"><spring:message code="language.en"
/></a>
   <br><br>
    <spring:message code="first"/><br><br>
    <a href="my/second"><spring:message code="second"/></a>
</body>
</html>
```

second.jsp 的代码如下：

```
<%@ page language="java" contentType="text/html; charset=UTF-8"
pageEncoding="UTF-8"%>
<%@taglib prefix="spring" uri="http://www.springframework.org/tags" %>
<%
```

```
String path = request.getContextPath();
String basePath=request.getScheme()+"://"+request.getServerName()+":"+
request.getServerPort()+path+"/";
%>
<!DOCTYPE html PUBLIC "-//W3C//DTD HTML 4.01 Transitional//EN"
"http://www.w3.org/TR/html4/loose.dtd">
<html>
<head>
<base href="<%=basePath%>">
<meta http-equiv="Content-Type" content="text/html; charset=UTF-8">
<title><spring:message code="second"/></title>
</head>
<body>
    <spring:message code="second"/><br><br>
    <a href="my/third"><spring:message code="third" arguments="888,999"
/></a>
</body>
</html>
```

third.jsp 的代码如下：

```
<%@ page language="java" contentType="text/html; charset=UTF-8" pageEncoding
="UTF-8"%>
<%@taglib prefix="spring" uri="http://www.springframework.org/tags" %>
<%
String path = request.getContextPath();
String basePath=request.getScheme()+"://"+request.getServerName()+":"+
request.getServerPort()+path+"/";
%>
<!DOCTYPE html PUBLIC "-//W3C//DTD HTML 4.01 Transitional//EN"
"http://www.w3.org/TR/html4/loose.dtd">
<html>
<head>
<base href="<%=basePath%>">
<meta http-equiv="Content-Type" content="text/html; charset=UTF-8">
<title><spring:message code="third"/></title>
</head>
<body>
    <spring:message code="third" arguments="888,999" /><br><br>
    <a href="my/first"><spring:message code="first"/></a>
</body>
</html>
```

3. 编写控制器类

该应用有两个控制器类，一个是 I18NTestController 处理语言种类选择请求，一个是 MyController 进行页面导航。

I18NTestController.java 的代码如下：

```
package controller;
import java.util.Locale;
import org.springframework.stereotype.Controller;
import org.springframework.web.bind.annotation.RequestMapping;
@Controller
public class I18NTestController {
    @RequestMapping("/i18nTest")
    /**
     * locale 接收请求参数 locale 值，并存储到 session 中
     */
    public String first(Locale locale){
        return "first";
    }
}
```

MyController 的代码如下：

```
package controller;
import org.springframework.stereotype.Controller;
import org.springframework.web.bind.annotation.RequestMapping;
@Controller
@RequestMapping("/my")
public class MyController {
    @RequestMapping("/first")
    public String first(){
        return "first";
    }
    @RequestMapping("/second")
    public String second(){
        return "second";
    }
    @RequestMapping("/third")
    public String third(){
        return "third";
    }
}
```

4. 编写配置文件

配置文件 springmvc-servlet.xml 的代码如下：

```
<?xml version="1.0" encoding="UTF-8"?>
<beans xmlns="http://www.springframework.org/schema/beans"
    xmlns:xsi="http://www.w3.org/2001/XMLSchema-instance"
    xmlns:p="http://www.springframework.org/schema/p"
```

```
    xmlns:context="http://www.springframework.org/schema/context"
    xmlns:mvc="http://www.springframework.org/schema/mvc"
    xsi:schemaLocation="
    http://www.springframework.org/schema/beans
    http://www.springframework.org/schema/beans/spring-beans.xsd
       http://www.springframework.org/schema/context
       http://www.springframework.org/schema/context/spring-context.xsd
       http://www.springframework.org/schema/mvc
       http://www.springframework.org/schema/mvc/spring-mvc.xsd">
    <!-- 使用扫描机制，扫描包 -->
    <context:component-scan base-package="controller" />
    <context:annotation-config />  <!-- 激活 Bean 中定义的注解 -->
    <mvc:annotation-driven />
    <!-- 配置视图解析器 -->
    <bean
        class="org.springframework.web.servlet.view.InternalResource_
ViewResolver"
        id="internalResourceViewResolver">
        <!-- 前缀 -->
        <property name="prefix" value="/WEB-INF/jsp/" />
        <!-- 后缀 -->
        <property name="suffix" value=".jsp" />
    </bean>
    <!-- 国际化操作拦截器 如果采用基于（Session/Cookie）则必须配置 -->
    <mvc:interceptors>
       <bean class="org.springframework.web.servlet.i18n.LocaleChange_
Interceptor"/>
    </mvc:interceptors>
    <!-- 存储区域设置信息 -->
    <bean id="localeResolver"
       class="org.springframework.web.servlet.i18n.
SessionLocaleResolver" >
       <property name="defaultLocale" value="zh_CN"></property>
     </bean>
    <!-- 加载国际化资源文件 -->
    <bean id="messageSource" class="org.springframework.context.support
.ReloadableResourceBundleMessageSource">
        <!-- <property name="basename" value="classpath:messages" /> -->
        <property name="basename" value="/WEB-INF/resource/messages" />
    </bean>
</beans>
```

6.4 本章小结

本章主要讲解 Spring MVC 的国际化知识。本章首先详细讲述了国际化资源文件的加载方式、语言区域选择、国际化信息显示，最后给出了一个让用户自行选择语言的示例，介绍了 Spring MVC 国际化的内在原理。

习题 6

1．在 JSP 页面中可以通过 Spring 提供的（　　）标签来输出国际化信息。

A．input　　B．message　　C．submit　　D．text

2．资源文件的后缀名为（　　）。

A．txt　　B．doc　　C．property　　D．properties

3．什么是国际化？国际化资源文件的命名格式是什么？

第7章

文件的上传与下载

学习目的与要求

本章重点讲解如何使用 Spring MVC 框架进行文件的上传与下载。通过本章的学习，读者应掌握 Spring MVC 框架的单文件上传、多文件上传以及文件下载。

本章主要内容

- 单文件上传
- 多文件上传
- 文件下载

文件上传是 Web 应用经常需要面对的问题。对于 Java 应用而言，上传文件有多种方式，包括使用文件流手工编程上传、基于 commons-fileupload 组件的文件上传、基于 Servlet 3 及以上版本的文件上传等。后两种方式在作者的另一本教材（《基于 Eclipse 平台的 JSP 应用教程》）中已经阐述。本章将重点介绍如何使用 Spring MVC 框架进行文件上传。

7.1 文件上传

Spring MVC 框架的文件上传是基于 commons-fileupload 组件的文件上传，只不过 Spring MVC 框架在原有文件上传组件上做了进一步封装，简化了文件上传的代码实现，取消了不同上传组件上的编程差异。

7.1.1 commons-fileupload 组件

由于 Spring MVC 框架的文件上传是基于 commons-fileupload 组件的文件上传，因此，需要将与 commons-fileupload 组件相关的 jar（commons-fileupload-1.3.1.jar 和 commons-io-2.4.jar）复制到 Spring MVC 应用的 WEB-INF/lib 目录下。下面讲解一下如何下载相关 jar 包。

Commons 是 Apache 开放源代码组织中的一个 Java 子项目，该项目包括文件上传、命令行处理、数据库连接池、XML 配置文件处理等模块。fileupload 就是其中用来处理基于表单的文件上传的子项目，commons-fileupload 组件性能优良，并支持任意大小文件的上传。

commons-fileupload 组件可以从 http://commons.apache.org/proper/commons-fileupload/ 下载，本书采用的版本是 1.3.1。下载它的 Binaries 压缩包（commons-fileupload-1.3.1-bin.zip），解压后的目录中有两个子目录，分别是 lib 和 site。lib 目录下有个 JAR 文件 commons-fileupload-1.3.1.jar，该文件是 commons-fileupload 组件的类库。site 目录中是 commons-fileupload 组件的文档，也包括 API 文档。

commons-fileupload 组件依赖于 Apache 的另外一个项目 commons-io，该组件可以从 http://commons.apache.org/proper/commons-io/下载，本书采用的版本是 2.4。下载它的 Binaries 压缩包（commons-io-2.4-bin.zip），解压缩后的目录中有 4 个 JAR 文件，其中有一个 commons-io-2.4.jar 文件，该文件是 commons-io 的类库。

7.1.2 基于表单的文件上传

标签<input type="file"/>在浏览器中会显示一个输入框和一个按钮，输入框可供用户填写本地文件的文件名和路径，按钮可以让浏览器打开一个文件选择框供用户选择文件。

文件上传的表单例子如下：

```
<form action="upload" method="post" enctype="multipart/form-data">
      <input type="file" name="myfile"/>
      ⋮
</form>
```

使用基于表单的文件上传时，不要忘记使用 enctype 属性，并将它的值设置为 multipart/form-data。同时，表单的提交方式应设置为 post。为什么需要这样呢？下面从 enctype 属性说起。

表单的 enctype 属性指定的是表单数据的编码方式，该属性有如下三个值：

- application/x-www-form-urlencoded：这是默认的编码方式，它只处理表单域里的 value 属性值。
- multipart/form-data：这种编码方式会以二进制流的方式来处理表单数据，会将文件域指定文件的内容也封装到请求参数里。
- text/plain：这种编码方式仅当表单的 action 属性为 mailto:URL 的形式时才使用，主要适用于直接通过表单发送邮件。

由上面三个属性值的解释可知，基于表单上传文件时，enctype 的属性值应为 multipart/form-data。

7.1.3 MultipartFile 接口

在 Spring MVC 框架中，上传文件时，文件的相关信息及操作会封装到 MultipartFile

对象中。因此，开发者只需要使用 MultipartFile 类型声明模型类的一个属性，即可以对被上传文件进行操作。该接口具有如下方法。

- byte[] getBytes()：以字节数组的形式返回文件的内容。
- String getContentType()：返回文件的内容类型。
- InputStream getInputStream()：返回一个 InputStream，从中可读取文件的内容。
- String getName()：返回请求参数的名称。
- String getOriginalFilename()：返回客户端提交的原始文件名称。
- long getSize()：返回文件的大小，单位为字节。
- boolean isEmpty()：判断被上传文件是否为空。
- void transferTo(File destination)：将上传文件保存到目标目录下。

上传文件时，需要在配置文件中使用 spring 的 CommosMultipartResolver 配置 MultipartResolver 用于文件上传。下面从单文件上传开始讲解该接口的使用方法。

7.1.4 单文件上传

本节通过一个示例讲解 Spring MVC 框架如何实现单文件的上传。假设有如图 7.1 所示的文件上传页面，为了完成文件上传，应该将该页面表单的 enctype 属性设置为 multipart/form-data。该页面（oneFile.jsp）的代码如下：

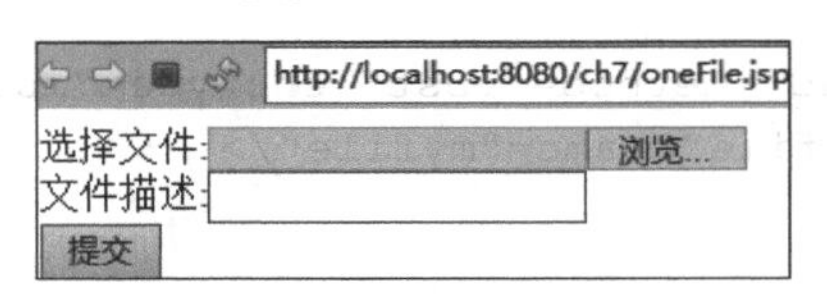

图 7.1 文件上传页面

```
<%@ page language="java" contentType="text/html; charset=UTF-8" pageEncoding
="UTF-8"%>
<%
String path = request.getContextPath();
String basePath = request.getScheme()+"://"+request.getServerName()+ ":"
+request.getServerPort()+path+"/";
%>
<!DOCTYPE html PUBLIC "-//W3C//DTD HTML 4.01 Transitional//EN"
"http://www.w3.org/TR/html4/loose.dtd">
<html>
<head>
<base href="<%=basePath%>">
<meta http-equiv="Content-Type" content="text/html; charset=UTF-8">
<title>Insert title here</title>
</head>
<body>
<form action="onefile" method="post" enctype="multipart/form-data">
    选择文件:<input type="file" name="myfile"> <br>
```

```
    文件描述:<input type="text" name="description"> <br>
  <input type="submit" value="提交">
</form>
</body>
</html>
```

上传文件时，需要用模型类 FileDomain 封装文件信息，在该类使用 MultipartFile 类型声明属性 myfile。模型类具体代码如下：

```
package domain;
import org.springframework.web.multipart.MultipartFile;
public class FileDomain {
    private String description;
    private MultipartFile myfile;
    public String getDescription() {
        return description;
    }
    public void setDescription(String description) {
        this.description = description;
    }
    public MultipartFile getMyfile() {
        return myfile;
    }
    public void setMyfile(MultipartFile myfile) {
        this.myfile = myfile;
    }
}
```

当文件上传页面提交请求时，请求由 onefile 发送到控制器，然后由控制器类 FileUploadController 的 oneFileUpload 方法处理，FileUploadController 类的代码如下：

```
package controller;
import java.io.File;
import java.util.List;
import javax.servlet.http.HttpServletRequest;
import org.apache.commons.logging.Log;
import org.apache.commons.logging.LogFactory;
import org.springframework.stereotype.Controller;
import org.springframework.web.bind.annotation.ModelAttribute;
import org.springframework.web.bind.annotation.RequestMapping;
import org.springframework.web.multipart.MultipartFile;
import domain.FileDomain;
import domain.MultiFileDomain;
@Controller
public class FileUploadController {
    // 得到一个用来记录日志的对象，这样打印信息的时候能够标记打印的是哪个类的信息
    private static final Log logger = LogFactory.getLog (FileUpload_
```

```
Controller.class);
    /**
     * 单文件上传
     */
    @RequestMapping("/onefile")
    public String oneFileUpload(@ModelAttribute FileDomain fileDomain,
HttpServletRequest request){
        /*上传文件的保存位置"/uploadfiles"，该位置是指
        workspace\.metadata\.plugins\org.eclipse.wst.server.core\tmp0\
wtpwebapps，发布后使用*/
        String realpath = request.getServletContext().getRealPath
("uploadfiles");
        String fileName = fileDomain.getMyfile().getOriginalFilename();
        File targetFile = new File(realpath, fileName);
        if(!targetFile.exists()){
          targetFile.mkdirs();
        }
        //上传
       try {
        fileDomain.getMyfile().transferTo(targetFile);
        logger.info("成功");
       } catch (Exception e) {
          e.printStackTrace();
       }
       /*
            @ModelAttribute FileDomain fileDomain这就拥有这样一个功能:
            model.addAttribute("fileDomain",fileDomain)所以此处不需要
        */
        return "showOne";
    }
    /**
     * 多文件上传
     */
    @RequestMapping("/multifile")
    public  String  multiFileUpload(@ModelAttribute  MultiFileDomain
multiFileDomain, HttpServletRequest request){
        String realpath = request.getServletContext().getRealPath
("uploadfiles");
        //String realpath = "D:/spring mvc workspace/ch7/WebContent/
uploadfiles";
        File targetDir = new File(realpath);
        if(!targetDir.exists()){
            targetDir.mkdirs();
        }
        List<MultipartFile> files = multiFileDomain.getMyfile();
        for (int i = 0; i < files.size(); i++) {
```

```
                MultipartFile file = files.get(i);
                String fileName = file.getOriginalFilename();
                File targetFile = new File(realpath,fileName);
                //上传
               try {
                file.transferTo(targetFile);
               } catch (Exception e) {
                  e.printStackTrace();
               }
           }
           logger.info("成功");
           return "showMulti";
       }
}
```

文件上传成功后，显示页面 showOne.jsp 的效果如图 7.2 所示。showOne.jsp 的代码如下：

图 7.2　显示页面

```
<%@ page language="java" contentType="text/html; charset=UTF-8"
pageEncoding="UTF-8"%>
<%
String path = request.getContextPath();
String basePath = request.getScheme()+"://"+request.getServerName()+":"+
request.getServerPort()+path+"/";
%>
<!DOCTYPE html PUBLIC "-//W3C//DTD HTML 4.01 Transitional//EN"
"http://www.w3.org/TR/html4/loose.dtd">
<html>
<head>
<base href="<%=basePath%>">
<meta http-equiv="Content-Type" content="text/html; charset=UTF-8">
<title>Insert title here</title>
</head>
<body>
    ${fileDomain.description }<br>
    <!-- fileDomain.getMyfile().getOriginalFilename() -->
    ${fileDomain.myfile.originalFilename }
</body>
</html>
```

上传文件时，需要在配置文件中使用 spring 的 CommosMultipartResolver 配置

MultipartResolver 用于文件上传。应用的配置文件 springmvc-servlet.xml 的代码如下：

```
<?xml version="1.0" encoding="UTF-8"?>
<beans xmlns="http://www.springframework.org/schema/beans"
    xmlns:xsi="http://www.w3.org/2001/XMLSchema-instance"
    xmlns:p="http://www.springframework.org/schema/p"
    xmlns:context="http://www.springframework.org/schema/context"
    xmlns:mvc="http://www.springframework.org/schema/mvc"
    xsi:schemaLocation="
    http://www.springframework.org/schema/beans
    http://www.springframework.org/schema/beans/spring-beans.xsd
       http://www.springframework.org/schema/context
       http://www.springframework.org/schema/context/spring-context.xsd
       http://www.springframework.org/schema/mvc
       http://www.springframework.org/schema/mvc/spring-mvc.xsd">
    <!-- 使用扫描机制，扫描包 -->
    <context:component-scan base-package="controller" />
    <mvc:annotation-driven />
    <!-- 配置视图解析器 -->
    <bean
        class="org.springframework.web.servlet.view.InternalResource_
ViewResolver"
        id="internalResourceViewResolver">
        <!-- 前缀 -->
        <property name="prefix" value="/WEB-INF/jsp/" />
        <!-- 后缀 -->
        <property name="suffix" value=".jsp" />
    </bean>
    <!-- 配置 MultipartResolver 用于文件上传 使用 spring 的 CommosMultipart_
Resolver -->
   <bean id="multipartResolver" class="org.springframework.web. multipart
.commons.CommonsMultipartResolver"
       p:defaultEncoding="UTF-8"
       p:maxUploadSize="5400000"
       p:uploadTempDir="fileUpload/temp"
    >
     <!--D:\spring mvc workspace\.metadata\.plugins\org.eclipse.wst
.server.core\tmp0\wtpwebapps\fileUpload  -->
   </bean>
   <!-- defaultEncoding="UTF-8" 是请求的编码格式，默认为 iso-8859-1
   maxUploadSize="5400000" 是允许上传文件的最大值，单位为字节
   uploadTempDir="fileUpload/temp" 为上传文件的临时路径 -->
</beans>
```

为防止中文乱码，需要在 web.xml 文件中添加字符编码过滤器，这里不再赘述。

最后，通过地址 http://localhost:8080/ch7/oneFile.jsp 测试单文件上传。

7.1.5 多文件上传

本节通过一个示例讲解 Spring MVC 框架如何实现多文件上传。假设有如图 7.3 所示的文件上传页面，为了完成文件上传，应该将该页面表单的 enctype 属性设置为 multipart/form-data。该页面（multiFiles.jsp）的代码如下：

图 7.3 多文件上传选择页面

```
<%@ page language="java" contentType="text/html; charset=UTF-8"
pageEncoding="UTF-8"%>
<%
String path = request.getContextPath();
String basePath = request.getScheme()+"://"+request.getServerName()+":"+
request.getServerPort()+path+"/";
%>
<!DOCTYPE html PUBLIC "-//W3C//DTD HTML 4.01 Transitional//EN"
"http://www.w3.org/TR/html4/loose.dtd">
<html>
<head>
<base href="<%=basePath%>">
<meta http-equiv="Content-Type" content="text/html; charset=UTF-8">
<title>Insert title here</title>
</head>
<body>
<form action="multifile" method="post" enctype="multipart/form-data">
    选择文件 1:<input type="file" name="myfile">  <br>
    文件描述 1:<input type="text" name="description"> <br>
    选择文件 2:<input type="file" name="myfile">  <br>
    文件描述 2:<input type="text" name="description"> <br>
    选择文件 3:<input type="file" name="myfile">  <br>
    文件描述 3:<input type="text" name="description"> <br>
 <input type="submit" value="提交">
</form>
</body>
</html>
```

上传多文件时，需要用模型类 MultiFileDomain 封装文件信息，模型类具体代码如下：

```
package domain;
import java.util.List;
import org.springframework.web.multipart.MultipartFile;
public class MultiFileDomain {
    private List<String> description;
    private List<MultipartFile> myfile;
    public List<String> getDescription() {
        return description;
    }
    public void setDescription(List<String> description) {
        this.description = description;
    }
    public List<MultipartFile> getMyfile() {
        return myfile;
    }
    public void setMyfile(List<MultipartFile> myfile) {
        this.myfile = myfile;
    }
}
```

当多文件上传页面提交请求时，请求由 multifile 发送到控制器，然后由控制器类 FileUploadController 的 multiFileUpload 方法处理。FileUploadController 类的代码见第 7.1.4 节。

文件上传成功后，显示页面 showMulti.jsp 的效果如图 7.4 所示。showMulti.jsp 的代码如下：

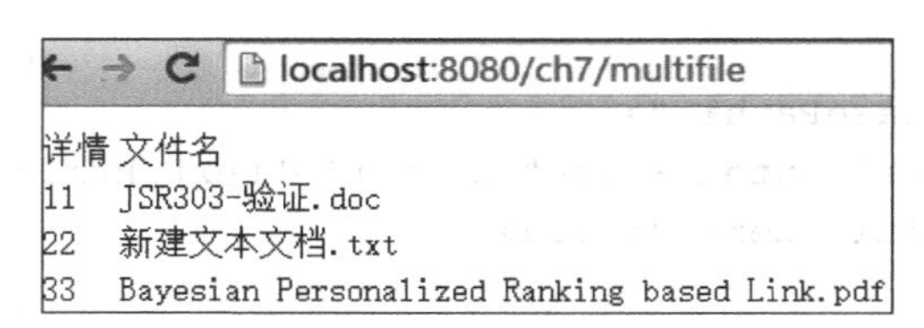

图 7.4 显示页面

```
<%@ page language="java" contentType="text/html; charset=UTF-8" pageEncoding
="UTF-8"%>
<%@ taglib uri="http://java.sun.com/jsp/jstl/core" prefix="c" %>
<%
String path = request.getContextPath();
String basePath = request.getScheme()+"://"+request.getServerName()+":"+
request.getServerPort()+path+"/";
%>
<!DOCTYPE html PUBLIC "-//W3C//DTD HTML 4.01 Transitional//EN"
"http://www.w3.org/TR/html4/loose.dtd">
<html>
<head>
<base href="<%=basePath%>">
```

```
<meta http-equiv="Content-Type" content="text/html; charset=UTF-8">
<title>Insert title here</title>
</head>
<body>
    <table>
        <tr>
            <td>详情</td><td>文件名</td>
        </tr>
        <!-- 同时取两个数组的元素 -->
        <c:forEach items="${multiFileDomain.description}" var="description"
varStatus="loop">
            <tr>
                <td>${description}</td>
                <td>${multiFileDomain.myfile[loop.count-1].original_
Filename}</td>
            </tr>
        </c:forEach>
        <!-- fileDomain.getMyfile().getOriginalFilename() -->
    </table>
</body>
</html>
```

配置文件与 web.xml 的内容与第 7.1.4 节的相同。

最后，通过地址 http://localhost:8080/ch7/multiFiles.jsp 测试多文件上传。

7.1.6　实践环节

对第 7.1.5 节多文件上传实例的文件类型做一下限定，例如，只允许上传.bmp、.gif、.jpg、.ico 等类型的文件。

7.2　文 件 下 载

7.2.1　文件下载的实现方法

实现文件下载经常有两种方法：一是通过超链接实现下载，一是利用程序编码实现下载。超链接实现下载固然简单，但会暴露下载文件的真实位置，并且只能下载存放在 Web 应用程序所在的目录下的文件。利用程序编码实现下载，可以增加安全访问控制，还可以从任意位置提供下载的数据；可以将文件存放到 Web 应用程序以外的目录中，也可以将文件保存到数据库中。

利用程序实现下载需要设置两个报头：

（1）Web 服务器需要告诉浏览器其所输出内容的类型不是普通文本文件或 HTML

文件，而是一个要保存到本地的下载文件。设置 Content-Type 的值为 application/x-msdownload。

（2）Web 服务器希望浏览器不直接处理相应的实体内容，而是由用户选择将相应的实体内容保存到一个文件中，这需要设置 Content-Disposition 报头。该报头指定了接收程序处理数据内容的方式，在 HTTP 应用中只有 attachment 是标准方式，attachment 表示要求用户干预。在 attachment 后面还可以指定 filename 参数，该参数是服务器建议浏览器将实体内容保存到文件中的文件名称。

设置报头的示例如下：

```
response.setHeader("Content-Type", "application/x-msdownload" );
response.setHeader("Content-Disposition", "attachment; filename=" +
filename);
```

7.2.2 文件下载过程

下面通过一个实例讲述利用程序实现下载的过程。该实例要求从第 7.1 节上传文件的目录（workspace\.metadata\.plugins\org.eclipse.wst.server.core\tmp0\wtpwebapps\ch7\uploadfiles）中下载文件，具体开发步骤如下：

首先编写控制器类 FileDownController，该类中有三个方法：show、down 和 toUTF8String。show 方法获取被下载的文件名称；down 方法执行下载功能；toUTF8String 方法是下载保存时中文文件名字符编码转换方法。FileDownController 类的代码如下：

```
package controller;
import java.io.File;
import java.io.FileInputStream;
import java.io.UnsupportedEncodingException;
import java.util.ArrayList;
import javax.servlet.ServletOutputStream;
import javax.servlet.http.HttpServletRequest;
import javax.servlet.http.HttpServletResponse;
import org.apache.commons.logging.Log;
import org.apache.commons.logging.LogFactory;
import org.springframework.stereotype.Controller;
import org.springframework.ui.Model;
import org.springframework.web.bind.annotation.RequestMapping;
import org.springframework.web.bind.annotation.RequestParam;
@Controller
public class FileDownController {
    // 得到一个用来记录日志的对象，这样打印信息的时候能够标记打印的是哪个类的信息
    private static final Log logger = LogFactory.getLog(FileUpload_
Controller.class);
    /**
     * 显示要下载的文件
```

```
     */
    @RequestMapping("showDownFiles")
    public String show(HttpServletRequest request, Model model){
        //从 workspace\.metadata\.plugins\org.eclipse.wst.server.core\
tmp0\wtpwebapps\ch7\下载
        String  realpath  =  request.getServletContext().getRealPath
("uploadfiles");
        File dir = new File(realpath);
        File files[] = dir.listFiles();
        //获取该目录下的所有文件名
        ArrayList<String> fileName = new ArrayList<String>();
        for (int i = 0; i < files.length; i++) {
            fileName.add(files[i].getName());
        }
        model.addAttribute("files", fileName);
        return "showDownFiles";
    }
    /**
     * 执行下载
     */
    @RequestMapping("down")
    public String down(@RequestParam String filename, HttpServletRequest
request, HttpServletResponse response){
        String aFilePath = null;         //要下载的文件路径
        FileInputStream in = null;       //输入流
        ServletOutputStream out = null; //输出流
        try {
            //从 workspace\.metadata\.plugins\org.eclipse.wst.server.
core\tmp0\wtpwebapps 下载
            aFilePath = request.getServletContext().getRealPath
("uploadfiles");
            //设置下载文件使用的报头
            response.setHeader("Content-Type","application/x-msdownload");
            response.setHeader("Content-Disposition", "attachment;
                                filename="+ toUTF8String(filename));
            //读入文件
            in = new FileInputStream(aFilePath + "\\"+ filename);
            //得到响应对象的输出流，用于向客户端输出二进制数据
            out = response.getOutputStream();
            out.flush();
            int aRead = 0;
            byte b[] = new byte[1024];
            while ((aRead = in.read(b)) != -1 & in != null) {
                out.write(b,0,aRead);
            }
            out.flush();
```

```
            in.close();
            out.close();
        } catch (Throwable e) {
            e.printStackTrace();
        }
        logger.info("下载成功");
        return null;
    }
    /**
     * 下载保存时中文文件名字符编码转换方法
     */
    public  String toUTF8String(String str){
        StringBuffer sb = new StringBuffer();
        int len = str.length();
        for(int i = 0; i < len; i++){
            //取出字符中的每个字符
            char c = str.charAt(i);
            //Unicode 码值在 0~255 之间，不作处理
            if(c >= 0 && c <= 255){
                sb.append(c);
            }else{//转换 UTF-8 编码
                byte b[];
                try {
                    b = Character.toString(c).getBytes("UTF-8");
                } catch (UnsupportedEncodingException e) {
                    e.printStackTrace();
                    b = null;
                }
                //转换为%HH 的字符串形式
                for(int j = 0; j < b.length; j ++){
                    int k = b[j];
                    if(k < 0){
                        k &= 255;
                    }
                    sb.append("%" + Integer.toHexString(k).toUpperCase());
                }
            }
        }
        return sb.toString();
    }
}
```

下载文件示例需要一个显示被下载文件的 JSP 页面 showDownFiles.jsp，代码如下：

```
<%@  page  language="java"  contentType="text/html;  charset=UTF-8"
pageEncoding="UTF-8"%>
```

```
<%@ taglib uri="http://java.sun.com/jsp/jstl/core" prefix="c" %>
<%
String path = request.getContextPath();
String basePath = request.getScheme()+"://"+request.getServerName()+":"+
request.getServerPort()+path+"/";
%>
<!DOCTYPE html PUBLIC "-//W3C//DTD HTML 4.01 Transitional//EN"
"http://www.w3.org/TR/html4/loose.dtd">
<html>
<head>
<base href="<%=basePath%>">
<meta http-equiv="Content-Type" content="text/html; charset=UTF-8">
<title>Insert title here</title>
</head>
<body>
    <table>
        <tr>
            <td>被下载的文件名</td>
        </tr>
        <!-- 遍历 model 中的 files -->
        <c:forEach items="${files}" var="filename">
            <tr>
                <td><a href="down?filename=${filename}">${filename}</a>
</td>
            </tr>
        </c:forEach>
    </table>
</body>
</html>
```

配置文件与 web.xml 与第 7.1 节的相同，这里不再赘述。

最后，通过地址 http://localhost:8080/ch7/showDownFiles 测试下载示例。

7.3 本章小结

本章重点介绍了 Spring MVC 的文件上传，主要包括如何使用 MultipartFile 接口封装文件信息。最后介绍了如何利用文件流进行文件下载。

习 题 7

1．使用基于表单的文件上传时，应将表单的 enctype 属性值设置为（　　）。

A．multipart/form-data

B．application/x-www-form-urlencoded

C．text/plain

D．html/text

2．在 Spring MVC 框架中，如何限定上传文件的大小？

3．单文件上传与多文件上传有什么区别？

第 8 章

统一异常处理

学习目的与要求

本章重点讲解如何使用 Spring MVC 框架进行异常的统一处理。通过本章的学习，读者需要掌握 Spring MVC 框架统一异常处理的使用方法。

本章主要内容

- 简单异常处理 SimpleMappingExceptionResolver
- 实现 HandlerExceptionResolver 接口自定义异常
- 使用@ExceptionHandler 注解实现异常处理

在 Spring MVC 应用的开发中，不管是对底层数据库操作，还是业务层操作，还是控制层操作，都不可避免遇到各种可预知的、不可预知的异常需要处理。如果每个过程都单独处理异常，那么系统的代码耦合度高，工作量大且不好统一，以后维护的工作量也很大。

如果能将所有类型的异常处理从各层中解耦出来，这样既保证了相关处理过程的功能较单一，也实现了异常信息的统一处理和维护。幸运的是，Spring MVC 框架支持这样的实现。本章将从使用 Spring MVC 提供的简单异常处理器 SimpleMappingExceptionResolver、实现 Spring 的异常处理接口 HandlerExceptionResolver 自定义自己的异常处理器、使用@ExceptionHandler 注解实现异常处理这三种处理方式讲解 Spring MVC 应用的异常统一处理。

8.1 示例介绍

为了验证 Spring MVC 框架的三种异常处理方式的实际效果，需要开发一个测试应用 ch8，从 Dao 层、Service 层、Controller 层分别抛出不同的异常（SQLException、自定义异常和未知异常），然后分别集成三种方式进行异常处理，进而比较三种方式的优缺点。ch8 项目结构如图 8.1 所示。

三种异常处理方式的共通部分如下：Dao 层、Service 层、View 层、MyException、TestExceptionController 以及 web.xml。下面分别介绍这些共通部分。

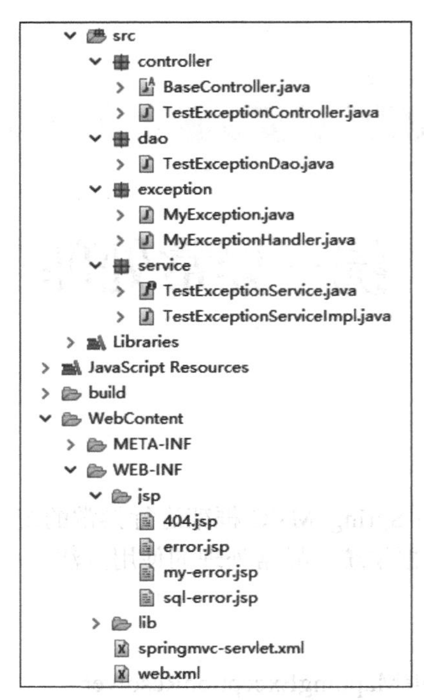

图 8.1　ch8 项目结构

1. Dao 层

TestExceptionDao.java 的代码如下：

```
package dao;
import java.sql.SQLException;
import org.springframework.stereotype.Repository;
import exception.MyException;
@Repository("testExceptionDao")
public class TestExceptionDao {
    public void daodb() throws Exception {
        throw new SQLException("Dao 中数据库异常");
    }
    public void daomy() throws Exception {
        throw new MyException("Dao 中自定义异常");
    }
    public void daono() throws Exception {
        throw new Exception("Dao 中未知异常");
    }
}
```

2. Service 层

TestExceptionService.java 的代码如下：

```
package service;
public interface TestExceptionService {
    public void servicemy() throws Exception;
    public void servicedb() throws Exception;
    public void daomy() throws Exception;
    public void daodb() throws Exception;
    public void serviceno() throws Exception;
    public void daono() throws Exception;
}
```

TestExceptionServiceImpl.java 的代码如下：

```
package service;
import java.sql.SQLException;
import org.springframework.beans.factory.annotation.Autowired;
import org.springframework.stereotype.Service;
import dao.TestExceptionDao;
import exception.MyException;
@Service("testExceptionService")
public class TestExceptionServiceImpl implements TestExceptionService{
    @Autowired
    private TestExceptionDao testExceptionDao;
    @Override
    public void servicemy() throws Exception {
        throw new MyException("Service 中自定义异常");
    }
    @Override
    public void servicedb() throws Exception {
        throw new SQLException("Service 中数据库异常");
    }
    @Override
    public void serviceno() throws Exception {
        throw new Exception("Service 中未知异常");
    }
    @Override
    public void daomy() throws Exception {
        testExceptionDao.daomy();
    }
    @Override
    public void daodb() throws Exception {
        testExceptionDao.daodb();
    }
    public void daono() throws Exception{
        testExceptionDao.daono();
    }
}
```

3. TestExceptionController

TestExceptionController.java 的代码如下：

```
package controller;
import java.sql.SQLException;
import org.springframework.beans.factory.annotation.Autowired;
import org.springframework.stereotype.Controller;
import org.springframework.web.bind.annotation.RequestMapping;
import org.springframework.web.bind.annotation.RequestMethod;
import exception.MyException;
import service.TestExceptionService;
@Controller
public class TestExceptionController{
    @Autowired
    private TestExceptionService testExceptionService;
    @RequestMapping(value = "/db", method = RequestMethod.GET)
    public void db() throws Exception {
        throw new SQLException("控制器中数据库异常");
    }
    @RequestMapping(value = "/my", method = RequestMethod.GET)
    public void my() throws Exception {
        throw new MyException("控制器中自定义异常");
    }
    @RequestMapping(value = "/no", method = RequestMethod.GET)
    public void no() throws Exception {
        throw new Exception("控制器中未知异常");
    }
    @RequestMapping(value = "/servicedb", method = RequestMethod.GET)
    public void servicedb() throws Exception {
        testExceptionService.servicedb();;
    }
    @RequestMapping(value = "/servicemy", method = RequestMethod.GET)
    public void servicemy() throws Exception {
        testExceptionService.servicemy();
    }
    @RequestMapping(value = "/serviceno", method = RequestMethod.GET)
    public void serviceno() throws Exception {
        testExceptionService.serviceno();
    }
    @RequestMapping(value = "/daodb", method = RequestMethod.GET)
    public void daodb() throws Exception {
        testExceptionService.daodb();
    }
    @RequestMapping(value = "/daomy", method = RequestMethod.GET)
    public void daomy() throws Exception {
```

```
        testExceptionService.daomy();
    }
    @RequestMapping(value = "/daono", method = RequestMethod.GET)
    public void daono() throws Exception {
        testExceptionService.daono();
    }
}
```

4. MyException

在测试示例中，定义了一个异常类，具体代码如下：

```
package exception;
public class MyException extends Exception {
    private static final long serialVersionUID = 1L;
    public MyException() {
        super();
    }
    public MyException(String message) {
        super(message);
    }
}
```

5. View 层

View 层共有 5 个 JSP 页面，下面分别介绍。
测试应用首页面 index.jsp 的代码如下：

```
<%@ page language="java" contentType="text/html; charset=UTF-8"
pageEncoding="UTF-8"%>
<%
String path = request.getContextPath();
String basePath=request.getScheme()+"://"+request.getServerName()+":"+
request.getServerPort()+path+"/";
%>
<!DOCTYPE html PUBLIC "-//W3C//DTD HTML 4.01 Transitional//EN"
"http://www.w3.org/TR/html4/loose.dtd">
<html>
<head>
<base href="<%=basePath%>">
<meta http-equiv="Content-Type" content="text/html; charset=UTF-8">
<title>Insert title here</title>
</head>
<body>
<h1>所有的演示例子</h1>
<h3><a href="daodb">处理 dao 中数据库异常</a></h3>
<h3><a href="daomy">处理 dao 中自定义异常</a></h3>
```

```
<h3><a href="daono">处理 dao 未知错误</a></h3>
<hr>
<h3><a href="servicedb">处理 service 中数据库异常</a></h3>
<h3><a href="servicemy">处理 service 中自定义异常</a></h3>
<h3><a href="serviceno">处理 service 未知错误</a></h3>
<hr>
<h3><a href="db">处理 controller 中数据库异常</a></h3>
<h3><a href="my">处理 controller 中自定义异常</a></h3>
<h3><a href="no">处理 controller 未知错误</a></h3>
<hr>
<!-- web.xml 中配置 404 -->
<h3><a href="404">404 错误</a></h3>
</body>
</html>
```

404 错误对应页面 404.jsp 的代码如下：

```
<%@  page  language="java"  contentType="text/html;  charset=UTF-8"
pageEncoding="UTF-8" isErrorPage="true"%>
<%
String path = request.getContextPath();
String basePath=request.getScheme()+"://"+request.getServerName()+":"+
request.getServerPort()+path+"/";
%>
<!DOCTYPE  html  PUBLIC  "-//W3C//DTD  HTML  4.01  Transitional//EN"
"http://www.w3.org/TR/html4/loose.dtd">
<html>
<head>
<base href="<%=basePath%>">
<meta http-equiv="Content-Type" content="text/html; charset=UTF-8">
<title>Insert title here</title>
</head>
<body>
资源已不在。
</body>
</html>
```

未知异常对应页面 error.jsp 的代码如下：

```
<%@  page  language="java"  contentType="text/html;  charset=UTF-8"
pageEncoding="UTF-8" isErrorPage="true"%>
<%
String path = request.getContextPath();
String basePath=request.getScheme()+"://"+request.getServerName()+":"+
request.getServerPort()+path+"/";
%>
<!DOCTYPE  html  PUBLIC  "-//W3C//DTD  HTML  4.01  Transitional//EN"
```

```
"http://www.w3.org/TR/html4/loose.dtd">
<html>
<head>
<base href="<%=basePath%>">
<meta http-equiv="Content-Type" content="text/html; charset=UTF-8">
<title>Insert title here</title>
</head>
<body>
<H1>未知错误：</H1><%=exception%>
<H2>错误内容：</H2>
<%
    exception.printStackTrace(response.getWriter());
%>
</body>
</html>
```

自定义异常对应页面 my-error.jsp 的代码如下：

```
<%@ page language="java" contentType="text/html; charset=UTF-8"
pageEncoding="UTF-8" isErrorPage="true"%>
<%
String path = request.getContextPath();
String basePath = request.getScheme()+"://"+request.getServerName()+":"+
request.getServerPort()+path+"/";
%>
<!DOCTYPE html PUBLIC "-//W3C//DTD HTML 4.01 Transitional//EN"
"http://www.w3.org/TR/html4/loose.dtd">
<html>
<head>
<base href="<%=basePath%>">
<meta http-equiv="Content-Type" content="text/html; charset=UTF-8">
<title>Insert title here</title>
</head>
<body>
<H1>自定义异常错误：</H1><%=exception%>
<H2>错误内容：</H2>
<%
exception.printStackTrace(response.getWriter());
%>
</body>
</html>
```

SQL 异常对应页面 sql-error.jsp 的代码如下：

```
<%@ page language="java" contentType="text/html; charset=UTF-8"
pageEncoding="UTF-8" isErrorPage="true"%>
<%
```

```
String path = request.getContextPath();
String basePath = request.getScheme()+"://"+request.getServerName()+":"+
request.getServerPort()+path+"/";
%>
<!DOCTYPE html PUBLIC "-//W3C//DTD HTML 4.01 Transitional//EN"
"http://www.w3.org/TR/html4/loose.dtd">
<html>
<head>
<base href="<%=basePath%>">
<meta http-equiv="Content-Type" content="text/html; charset=UTF-8">
<title>Insert title here</title>
</head>
<body>
<H1>数据库异常错误：</H1><%=exception%>
<H2>错误内容：</H2>
<%
exception.printStackTrace(response.getWriter());
%>
</body>
</html>
```

6．web.xml

对于 Unchecked Exception 而言，由于代码不强制捕获，往往被忽略，如果运行期产生了 Unchecked Exception，而代码中又没有进行相应的捕获和处理，则可能不得不面对 404、500……等服务器内部错误提示页面。所以在 web.xml 文件中添加了全局异常 404 处理，具体代码如下：

```
<error-page>
  <error-code>404</error-code>
  <location>/WEB-INF/jsp/404.jsp</location>
</error-page>
```

从上述 Dao 层、Service 层以及 Controller 层的代码中可以看出，它们只管通过 throw 和 throws 语句抛出异常，并不处理。下面分别从三种方式统一处理这些异常。

8.2 SimpleMappingExceptionResolver 类

org.springframework.web.servlet.handler.SimpleMappingExceptionResolver 需要在配置文件中提前配置异常类和 View 的对应关系然后才能进行统一异常处理。配置文件 springmvc-servlet.xml 的具体代码如下：

```
<?xml version="1.0" encoding="UTF-8"?>
<beans xmlns="http://www.springframework.org/schema/beans"
```

```
    xmlns:xsi="http://www.w3.org/2001/XMLSchema-instance"
    xmlns:p="http://www.springframework.org/schema/p"
    xmlns:context="http://www.springframework.org/schema/context"
    xmlns:mvc="http://www.springframework.org/schema/mvc"
    xsi:schemaLocation="
    http://www.springframework.org/schema/beans
    http://www.springframework.org/schema/beans/spring-beans.xsd
       http://www.springframework.org/schema/context
       http://www.springframework.org/schema/context/spring-context.xsd
       http://www.springframework.org/schema/mvc
       http://www.springframework.org/schema/mvc/spring-mvc.xsd">
    <!-- 使用扫描机制，扫描包 -->
    <context:component-scan base-package="controller" />
    <context:component-scan base-package="service" />
    <context:component-scan base-package="dao" />
    <mvc:annotation-driven />
    <!-- 配置视图解析器 -->
    <bean
        class="org.springframework.web.servlet.view.InternalResource_
ViewResolver"
        id="internalResourceViewResolver">
        <!-- 前缀 -->
        <property name="prefix" value="/WEB-INF/jsp/" />
        <!-- 后缀 -->
        <property name="suffix" value=".jsp" />
    </bean>
    <!-- 配置 SimpleMappingExceptionResolver -->
    <bean class="org.springframework.web.servlet.handler.SimpleMapping_
ExceptionResolver">
        <!-- 定义默认的异常处理页面，当该异常类型注册时使用 -->
        <property name="defaultErrorView" value="error"></property>
        <!-- 定义异常处理页面用来获取异常信息的变量名，默认名为 exception -->
        <property name="exceptionAttribute" value="ex"></property>
        <!-- 定义需要特殊处理的异常，用类名或完全路径名作为 key，异常页名作为值 -->
        <property name="exceptionMappings">
            <props>
                <prop key="exception.MyException">my-error</prop>
                <prop key="java.sql.SQLException">sql-error</prop>
                <!-- 这里还可以继续扩展对不同异常类型的处理 -->
            </props>
        </property>
    </bean>
</beans>
```

配置完成后，就可以通过 SimpleMappingExceptionResolver 异常处理器统一处理第 8.1 节中的异常。

通过地址 http://localhost:8080/ch8/测试应用。

8.3 HandlerExceptionResolver 接口

org.springframework.web.servlet.HandlerExceptionResolver 接口用于解析请求处理过程中所产生的异常。开发者可以开发该接口的实现类进行 Spring MVC 应用的异常统一处理。ch8 应用中开发了一个该接口的实现类 MyExceptionHandler，具体代码如下：

```
package exception;
import java.sql.SQLException;
import java.util.HashMap;
import java.util.Map;
import javax.servlet.http.HttpServletRequest;
import javax.servlet.http.HttpServletResponse;
import org.springframework.web.servlet.HandlerExceptionResolver;
import org.springframework.web.servlet.ModelAndView;
public class MyExceptionHandler implements HandlerExceptionResolver {
    @Override
public ModelAndView resolveException(HttpServletRequest arg0,HttpServlet_
            Response arg1, Object arg2,Exception arg3) {
            Map<String, Object> model = new HashMap<String, Object>();
           model.put("ex", arg3);
           // 根据不同错误转向不同页面（统一处理）
           if(arg3 instanceof MyException) {
               return new ModelAndView("my-error", model);
           }else if(arg3 instanceof SQLException) {
               return new ModelAndView("sql-error", model);
           } else {
               return new ModelAndView("error", model);
           }
    }
}
```

需要将实现类 MyExceptionHandler 在配置文件中托管给 Spring MVC 框架才能进行异常的统一处理。配置代码为<bean class="exception.MyExceptionHandler"/>。

实现 HandlerExceptionResolver 接口统一处理异常时，配置文件的代码修改如下：

```
<?xml version="1.0" encoding="UTF-8"?>
<beans xmlns="http://www.springframework.org/schema/beans"
    xmlns:xsi="http://www.w3.org/2001/XMLSchema-instance"
    xmlns:p="http://www.springframework.org/schema/p"
    xmlns:context="http://www.springframework.org/schema/context"
    xmlns:mvc="http://www.springframework.org/schema/mvc"
    xsi:schemaLocation="
```

```
    http://www.springframework.org/schema/beans
    http://www.springframework.org/schema/beans/spring-beans.xsd
       http://www.springframework.org/schema/context
       http://www.springframework.org/schema/context/spring-context.xsd
       http://www.springframework.org/schema/mvc
       http://www.springframework.org/schema/mvc/spring-mvc.xsd">
    <!-- 使用扫描机制，扫描包 -->
    <context:component-scan base-package="controller" />
    <context:component-scan base-package="service" />
    <context:component-scan base-package="dao" />
    <mvc:annotation-driven />
    <!-- 配置视图解析器 -->
    <bean
    class="org.springframework.web.servlet.view.InternalResourceView
Resolver"
        id="internalResourceViewResolver">
        <!-- 前缀 -->
        <property name="prefix" value="/WEB-INF/jsp/" />
        <!-- 后缀 -->
        <property name="suffix" value=".jsp" />
    </bean>
    <!-- 托管 MyExceptionHandler -->
    <bean class="exception.MyExceptionHandler"/>
</beans>
```

最后通过地址 http://localhost:8080/ch8/测试应用。

8.4 @ExceptionHandler 注解

创建 BaseController 类，并在类中使用@ExceptionHandler 注解声明异常处理方法，具体代码如下：

```
package controller;
import java.sql.SQLException;
import javax.servlet.http.HttpServletRequest;
import org.springframework.web.bind.annotation.ExceptionHandler;
import exception.MyException;
public abstract class BaseController {
     /** 基于@ExceptionHandler 异常处理 */
    @ExceptionHandler
   public String exception(HttpServletRequest request, Exception ex)  {
         request.setAttribute("ex", ex);
       // 根据不同错误转向不同页面
       if(ex instanceof SQLException) {
        return "sql-error";
```

```
        }else if(ex instanceof MyException) {
         return "my-error";
        } else {
         return "error";
        }
    }
}
```

将所有需要异常处理的 Controller 都继承 BaseController 类，示例代码如下：

```
@Controller
public class TestExceptionController extends BaseController{
    …
}
```

使用@ExceptionHandler 注解声明统一处理异常时，不需要配置任何信息。此时，配置文件的代码修改如下：

```
<?xml version="1.0" encoding="UTF-8"?>
<beans xmlns="http://www.springframework.org/schema/beans"
    xmlns:xsi="http://www.w3.org/2001/XMLSchema-instance"
    xmlns:p="http://www.springframework.org/schema/p"
    xmlns:context="http://www.springframework.org/schema/context"
    xmlns:mvc="http://www.springframework.org/schema/mvc"
    xsi:schemaLocation="
    http://www.springframework.org/schema/beans
    http://www.springframework.org/schema/beans/spring-beans.xsd
       http://www.springframework.org/schema/context
       http://www.springframework.org/schema/context/spring-context.xsd
       http://www.springframework.org/schema/mvc
       http://www.springframework.org/schema/mvc/spring-mvc.xsd">
    <!-- 使用扫描机制，扫描包 -->
    <context:component-scan base-package="controller" />
    <context:component-scan base-package="service" />
    <context:component-scan base-package="dao" />
    <mvc:annotation-driven />
    <!-- 配置视图解析器 -->
    <bean
        class="org.springframework.web.servlet.view.InternalResource_
ViewResolver"
        id="internalResourceViewResolver">
        <!-- 前缀 -->
        <property name="prefix" value="/WEB-INF/jsp/" />
        <!-- 后缀 -->
        <property name="suffix" value=".jsp" />
    </bean>
</beans>
```

最后通过地址 http://localhost:8080/ch8/测试应用。

8.5 本章小结

本章重点介绍了 Spring MVC 框架应用程序的统一异常处理的三种方法。从上面的处理过程可知，使用@ExceptionHandler 注解实现异常处理，具有集成简单、可扩展性好（只需要将要异常处理的 Controller 类继承于 BaseController 即可）、不需要附加 Spring 配置等优点，但该方法对已有代码存在入侵性（需要修改已有代码，使相关类继承于 BaseController）。

习题 8

1．简述 Spring MVC 框架中统一异常处理的常用方式。
2．如何使用@ExceptionHandler 注解进行统一异常处理？

第9章 chapter 9

EL与JSTL

学习目的与要求

本章主要介绍表达式语言（Expression Language，EL）和 JSP 标准标签库（Java Server Pages Standard Tag Library，JSTL）的基本用法。通过本章的学习，掌握 EL 表达式语法，掌握 EL 隐含对象，了解什么是 JSTL，掌握 JSTL 的核心标签库。

本章主要内容

- EL
- JSTL

在 JSP 页面中可以使用 Java 代码来实现页面显示逻辑，但网页中夹杂着 HTML 与 Java 代码，给网页的设计与维护带来困难。可以使用 EL 来访问和处理应用程序的数据，也可以使用 JSTL 来替换网页中实现页面显示逻辑的 Java 代码。这样 JSP 页面就尽量减少了 Java 代码的使用，为以后的维护提供了方便。

9.1 表达式语言 EL

EL 是 JSP 2.0 规范中增加的，它的基本语法为：

```
${表达式}
```

EL 表达式类似于 JSP 表达式<%=表达式%>，EL 语句中的表达式值会被直接送到浏览器显示。通过 page 指令的 isELIgnored 属性来说明是否支持 EL 表达式。isELIgnored 属性值为 False 时，JSP 页面可以使用 EL 表达式；isELIgnored 属性值为 true 时，JSP 页面不能使用 EL 表达式。isELIgnored 属性值默认为 False。

9.1.1 基本语法

EL 的语法简单，使用方便。它以“${”开始，以“}”结束。

1. “[ ]” 与 “.” 运算符

EL 使用“[ ]”和“.”运算符来访问数据，主要使用 EL 获取对象的属性，包括获取 JavaBean 的属性值、获取数组中的元素以及获取集合对象中的元素。对于 null 值直接以空字符串显示，而不是 null，运算时也不会发生错误或空指针异常。所以在使用 EL 访问对象的属性时，不需判断对象是否为 null 对象。这样就为编写程序提供了方便。

（1）获取 JavaBean 的属性值

假设在 JSP 页面中有这样一句话：

```
<%=user.getAge ()%>
```

那么，可以使用 EL 获取 user 对象的属性 age，代码如下：

```
${user.age}
```

或

```
${user["age"]}
```

其中，点运算符前面为 JavaBean 的对象 user，后面为该对象的属性 age，表示利用 user 对象的 getAge()方法取值并显示在网页上。

（2）获取数组中的元素

假设在 Controller 或 Servlet 中有这样一段话：

```
String dogs[] = {"lili","huahua","guoguo"};
request.setAttribute("array", dogs);
```

那么，在对应视图 JSP 中可以使用 EL 取出数组中的元素（也可以使用第 8.2 节的 JSTL 遍历数组），代码如下：

```
${array[0]}
${array[1]}
${array[2]}
```

（3）获取集合对象中的元素

假设在 Controller 或 Servlet 中有这样一段话：

```
ArrayList<UserBean> users = new ArrayList<UserBean>();
UserBean ub1 = new UserBean("zhang",20);
UserBean ub2 = new UserBean("zhao",50);
users.add(ub1);
users.add(ub2);
request.setAttribute("array", users);
```

其中，UserBean有两个属性：name和age，那么在对应视图JSP页面中可以使用EL取出UserBean中的属性（也可以使用第8.2节的JSTL遍历数组），代码如下：

```
${array[0].name}  ${array[0].age}
${array[1].name}  ${array[1].age}
```

2. 算术运算符

在 EL 表达式中有 5 个算术运算符，如表 9.1 所示。

表 9.1　EL 的算术运算符

算术运算符	说　明	示　例	结　果
+	加	${13+2}	15
-	减	${13-2}	11
*	乘	${13*2}	26
/（或 div）	除	${13/2}或${13 div 2}	6.5
%（或 mod）	取模（求余）	${13%2}或${13 mod 2}	1

3. 关系运算符

在 EL 表达式中有 6 个关系运算符，如表 9.2 所示。

表 9.2　EL 的关系运算符

关系运算符	说　明	示　例	结　果
==（或 eq）	等于	${13 == 2}或${13 eq 2}	False
!=（或 ne）	不等于	${13 != 2}或${13 ne 2}	True
<（或 lt）	小于	${13 < 2}或${13 lt 2}	False
>（或 gt）	大于	${13 > 2}或${13 gt 2}	True
<=（或 le）	小于等于	${13 <= 2}或${13 le 2}	False
>=（或 ge）	大于等于	${13 >= 2}或${13 ge 2}	True

4. 逻辑运算符

在 EL 表达式中有 3 个逻辑运算符，如表 9.3 所示。

表 9.3　EL 的逻辑运算符

逻辑运算符	说　明	示　例	结　果
&&（或 and）	逻辑与	如果 A 为 True，B 为 False，则 A && B（或 A and B）	False
||（或 or）	逻辑或	如果 A 为 True，B 为 False，则 A || B（或 A or B）	True
!（或 not）	逻辑非	如果 A 为 True，则!A（或 not A）	False

5. empty 运算符

empty 运算符用于检测一个值是否为 null，例如，变量 A 不存在，则${empty A}返回的结果为 True。

6．条件运算符

EL 中的条件运算符是“? :”，例如，${A ? B:C}，如果 A 为 True，计算 B 并返回其结果，如果 A 为 False，计算 C 并返回其结果。

9.1.2 EL 隐含对象

EL 隐含对象共有 11 个，在本书中只是介绍几个常用的 EL 隐含对象：pageScope、requestScope、sessionScope、applicationScope、param 以及 paramValues。

1．与作用范围相关的隐含对象

与作用范围有关的 EL 隐含对象有 pageScope、requestScope、sessionScope 和 applicationScope，分别可以获取 JSP 隐含对象 pageContext、request、session 和 application 中的数据。如果在 EL 中没有使用隐含对象指定作用范围，则会依次从 page、request、session、application 范围查找，找到就直接返回，不再继续找下去，如果所有范围都没有找到，就返回空字符串。获取数据的格式如下：

```
${EL 隐含对象.关键字对象.属性}
```

或

```
${EL 隐含对象.关键字对象}
```

例如：

```
<jsp:useBean id="user" class="bean.UserBean" scope="page"/><!-- bean 标签-->
<jsp:setProperty name="user" property="name" value="EL 隐含对象" />
name: ${pageScope.user.name}
```

再比如，在 Controller 或 Servlet 中有这样一段话：

```
ArrayList<UserBean> users = new ArrayList<UserBean>();
UserBean ub1 = new UserBean("zhang",20);
UserBean ub2 = new UserBean("zhao",50);
users.add(ub1);
users.add(ub2);
request.setAttribute("array", users);
```

其中，UserBean 有两个属性：name 和 age，那么在对应视图 JSP 中，request 有效的范围内可以使用 EL 取出 UserBean 的属性（也可以使用第 8.2 节的 JSTL 遍历数组），代码如下：

```
${requestScope.array[0].name}  ${requestScope.array[0].age}
${requestScope.array[1].name}  ${requestScope.array[1].age}
```

2. 与请求参数相关的隐含对象

与请求参数相关的 EL 隐含对象有 param 和 paramValues。获取数据的格式如下：

```
${EL 隐含对象.参数名}
```

例如，input.jsp 的代码如下：

```
<form method = "post" action = "param.jsp">
    <p>姓名：<input type="text" name="username" size="15" /></p>
    <p>兴趣：
    <input type="checkbox" name="habit" value="看书"/>看书
    <input type="checkbox" name="habit" value="玩游戏"/>玩游戏
    <input type="checkbox" name="habit" value="旅游"/>旅游
    <p>
    <input type="submit" value="提交"/>
</form>
```

那么，在 param.jsp 页面中可以使用 EL 获取参数值，代码如下：

```
<%request.setCharacterEncoding("GBK");%>
<body>
<h2>EL 隐含对象 param、paramValues</h2>
姓名： ${param.username}</br>
兴趣：
${paramValues.habit[0]}
${paramValues.habit[1]}
${paramValues.habit[2]}
```

【例 9-1】 编写一个 Controller，在该控制器类处理方法中使用 request 对象和 Model 对象存储数据，然后从处理方法转发到 show.jsp 页面，在 show.jsp 页面中显示 request 对象的数据。首先，运行控制器的处理方法，在 IE 地址栏中输入“http://localhost:8080/ch9/input”。

程序运行结果如图 9.1 所示。

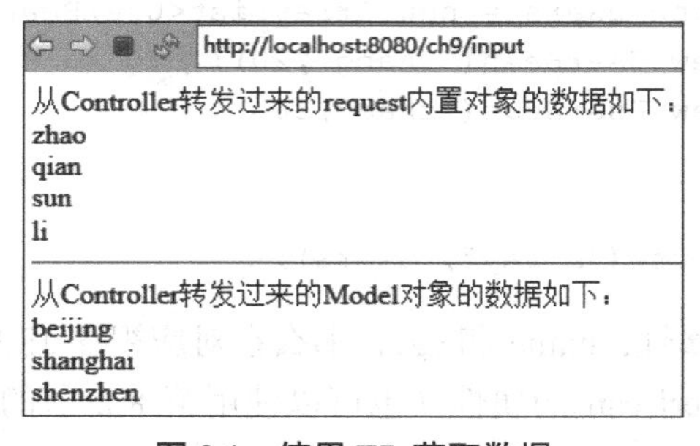

图 9.1 使用 EL 获取数据

InputController.java 的代码如下：

```
package controller;
import javax.servlet.http.HttpServletRequest;
import org.springframework.stereotype.Controller;
import org.springframework.ui.Model;
import org.springframework.web.bind.annotation.RequestMapping;
@Controller
public class InputController {
    @RequestMapping("input")
    public String input(HttpServletRequest request, Model model){
        String names[] = { "zhao", "qian", "sun", "li" };
        request.setAttribute("name", names);
        String address[] = { "beijing", "shanghai", "shenzhen"};
        model.addAttribute("address", address);
        return "show";
    }
}
```

例 9-1 页面文件 show.jsp 的代码如下：

```
<%@ page language="java" contentType="text/html; charset=UTF-8"
pageEncoding="UTF-8"%>
<%
String path = request.getContextPath();
String basePath = request.getScheme()+"://"+request.getServerName()+":"+
request.getServerPort()+path+"/";
%>
<!DOCTYPE html PUBLIC "-//W3C//DTD HTML 4.01 Transitional//EN"
"http://www.w3.org/TR/html4/loose.dtd">
<html>
<head>
<base href="<%=basePath%>">
<meta http-equiv="Content-Type" content="text/html; charset=UTF-8">
<title>Insert title here</title>
</head>
<body>
    从 Controller 转发过来的 request 内置对象的数据如下：<br>
    ${requestScope.name[0]}<br>
    ${requestScope.name[1]}<br>
    ${requestScope.name[2]}<br>
    ${requestScope.name[3]}<br>
    <hr>
    从 Controller 转发过来的 Model 对象的数据如下：<br>
    ${address[0]}<br>
    ${address[1]}<br>
    ${address[2]}<br>
</body>
```

```
</html>
```

9.1.3 实践环节

把例 9-1 的 Controller 中 input 方法的代码修改如下：

```
@RequestMapping("input")
public String input(HttpServletRequest request){
    Map<String, String> names = new HashMap<String, String>();
    names.put("first", "zhao");
    names.put("second", "qian");
    names.put("third", "sun");
    names.put("forth", "li");
    request.setAttribute("name", names);
    return "show";
}
```

请修改例 9-1 的 show.jsp，显示 map 中的数据。

9.2 JSP 标准标签库 JSTL

JSTL 规范由 Sun 公司制定，Apache 的 Jakarta 小组负责实现。JSTL 标准标签库由 5 个不同功能的标签库组成，包括 Core、I18N、XML、SQL 以及 Functions，本节只是简要介绍 JSTL 的 Core 和 Functions 标签库中几个常用的标签。

9.2.1 配置 JSTL

JSTL 现在已经是 Java EE5 的一个组成部分了，如果采用支持 Java EE5 或 Java EE6 的集成开发环境开发 Web 应用程序时，就不需要再配置 JSTL 了。但本书采用的是 Eclipse 平台，因此需要配置 JSTL。配置 JSTL 的步骤如下：

1. 复制 JSTL 的标准实现

在 Tomcat 的\webapps\examples\WEB-INF\lib 目录下，找到“taglibs-standard-impl-1.2.5.jar”和“taglibs-standard-spec-1.2.5.jar”文件，然后复制到 Web 工程的 WEB-INF\lib 目录下。

2. 使用 taglib 标记定义前缀与 uri 引用

如果使用 Core 标签库，首先需要在 JSP 页面中使用 taglib 标记定义前缀与 uri 引用，代码如下：

```
<%@ taglib prefix="c" uri="http://java.sun.com/jsp/jstl/core"%>
```

如果使用 Functions 标签库，首先需要在 JSP 页面中使用 taglib 标记定义前缀与 uri 引用，代码如下：

```
<%@ taglib prefix="fn" uri="http://java.sun.com/jsp/jstl/functions"%>
```

9.2.2 核心标签库之通用标签

1. <c:out>标签

<c:out>用来显示数据的内容，与 <%= 表达式 %> 或${表达式}类似。格式如下：

```
<c:out value="输出的内容" [default="defaultValue"]/>
```

或

```
<c:out value="输出的内容">
    defaultValue
</c:out>
```

其中，value 值可以是一个 EL 表达式，也可以是一个字符串；default 可有可无，当 value 值不存在时，就输出 defaultValue。例如：

```
<c:out value="${param.data}" default="没有数据" />
<br>
<c:out value="${param.nothing}" />
<br>
<c:out value="这是一个字符串" />
```

输出的结果如图 9.2 所示。

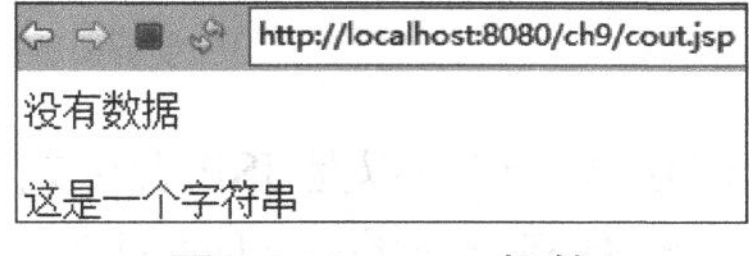

图 9.2　<c:out>标签

2. <c:set>标签

- 设置作用域变量

可以使用<c:set>在 page、request、session、application 等范围内设置一个变量。格式如下：

```
<c:set value="value" var="varName" [scope="page|request|session|
application"]/>
```

将 value 值赋值给变量 varName。例如：

```
<c:set value="zhao" var="userName" scope="session"/>
```

相当于

```
<%  session.setAttribute("userName","zhao"); %>
```

- 设置 JavaBean 的属性

使用<c:set>设置 JavaBean 的属性时，必须使用 target 属性进行设置。格式如下：

```
<c:set value="value" target="target"  property="propertyName"/>
```

将 value 赋值给 target 对象（JaveBean 对象）的 propertyName 属性。如果 target 为 null 或没有 set 方法则抛出异常。

3．<c:remove>标签

如果要删除某个变量，则可以使用<c:remove>标签。例如：

```
<c:remove var="userName" scope="session"/>
```

相当于

```
<%session.removeAttribute("userName") %>
```

9.2.3 核心标签库之流程控制标签

1．<c:if>标签

<c:if>标签实现 if 语句的作用，具体语法格式如下：

```
<c:if test="条件表达式">
    主体内容
</c:if>
```

其中，条件表达式可以是 EL 表达式，也可以是 JSP 表达式。如果表达式的值为 true，则会执行<c:if>的主体内容，但是没有相对应的<c:else>标签。如果想在条件成立时执行一块内容，不成立时执行另一块内容，则可以使用<c:choose>、<c:when>及<c:otherwise>标签。

2．<c:choose>、<c:when>及<c:otherwise>标签

<c:choose>、<c:when>及<c:otherwise>标签实现 if/elseif/else 语句的作用。具体语法格式如下：

```
<c:choose>
        <c:when test="条件表达式 1">
            主体内容 1
        </c:when>
        <c:when test="条件表达式 2">
```

```
            主体内容 2
        </c:when>
        <c:otherwise>
            表达式都不正确时，执行的主体内容
        </c:otherwise>
</c:choose>
```

【例 9-2】 编写一个 JSP 页面 ifelse.jsp，在该页面中使用<c:set>标签把两个字符串设置为 request 范围内的变量。使用<c:if>标签求出这两个字符串的最大值（按字典顺序比较大小），使用<c:choose>、<c:when>及<c:otherwise>标签求出这两个字符串的最小值。

例 9-2 页面文件 ifelse.jsp 的代码如下：

```
<%@ page language="java" contentType="text/html; charset=UTF-8"
pageEncoding="UTF-8"%>
<%@ taglib uri="http://java.sun.com/jsp/jstl/core" prefix="c" %>
<%
String path = request.getContextPath();
String basePath = request.getScheme()+"://"+request.getServerName()+":"+
request.getServerPort()+path+"/";
%>
<!DOCTYPE html PUBLIC "-//W3C//DTD HTML 4.01 Transitional//EN"
"http://www.w3.org/TR/html4/loose.dtd">
<html>
<head>
<base href="<%=basePath%>">
<meta http-equiv="Content-Type" content="text/html; charset=UTF-8">
<title>Insert title here</title>
</head>
<body>
    <c:set value="if" var="firstNumber" scope="request" />
    <c:set value="else" var="secondNumber" scope="request" />
    <c:if test="${firstNumber>secondNumber}">
        最大值为${firstNumber}
    </c:if>
    <c:if test="${firstNumber<secondNumber}">
        最大值为${secondNumber}
    </c:if>
    <c:choose>
        <c:when test="${firstNumber<secondNumber}">
            最小值为${firstNumber}
        </c:when>
        <c:otherwise>
            最小值为${secondNumber}
        </c:otherwise>
    </c:choose>
</body>
```

```
</html>
```

<c:when>及<c:otherwise>必须放在<c:choose>之中。当<c:when>的 test 结果为 true 时，会输出<c:when>的主体内容，而不理会<c:otherwise>的内容。<c:choose>中可有多个<c:when>，程序会从上到下进行条件判断，如果有个<c:when>的 test 结果为 true，就输出其主体内容，之后的<c:when>就不再执行。如果所有的<c:when>的 test 结果都为 false 时，则会输出<c:otherwise>的内容。<c:if>与<c:choose>也可以嵌套使用，例如：

```
<c:set value="fda" var="firstNumber" scope="request"/>
<c:set value="else" var="secondNumber" scope="request"/>
<c:set value="ddd" var="threeNumber" scope="request"/>
<c:if test="${firstNumber>secondNumber}">
    <c:choose>
        <c:when test="${firstNumber<threeNumber}">
            最大值为${threeNumber}
        </c:when>
        <c:otherwise>
            最大值为${firstNumber}
        </c:otherwise>
    </c:choose>
</c:if>
<c:if test="${secondNumber>firstNumber}">
    <c:choose>
        <c:when test="${secondNumber<threeNumber}">
            最大值为${threeNumber}
        </c:when>
        <c:otherwise>
            最大值为${secondNumber}
        </c:otherwise>
    </c:choose>
</c:if>
```

9.2.4 核心标签库之迭代标签

1. <c:forEach>标签

<c:forEach>标签可以实现程序中的 for 循环。语法格式如下：

```
<c:forEach var="变量名" items="数组或Collection对象">
    循环体
</c:forEach>
```

其中，items 属性可以是数组或 Collection 对象，每次循环读取对象中的一个元素，并赋值给 var 属性指定的变量，之后就可以在循环体使用 var 指定的变量获取对象的元素。例如，在 Controller 或 Servlet 中有这样一段代码：

```
ArrayList<UserBean> users = new ArrayList<UserBean>();
UserBean ub1 = new UserBean("zhao",20);
UserBean ub2 = new UserBean("qian",40);
UserBean ub3 = new UserBean("sun",60);
UserBean ub4 = new UserBean("li",80);
users.add(ub1);
users.add(ub2);
users.add(ub3);
users.add(ub4);
request.setAttribute("usersKey", users);
```

那么，在对应 JSP 页面中可以使用<c:forEach>循环遍历出数组中的元素。代码如下：

```
<table>
  <tr>
    <th>姓名</th>
    <th>年龄</th>
  </tr>
<c:forEach var="user" items="${requestScope.usersKey}">
  <tr>
    <td>${user.name}</td>
    <td>${user.age}</td>
  </tr>
</c:forEach>
</table>
```

有些情况下，需要为<c:forEach>标签指定 begin、end、step 和 varStatus 属性。begin 为迭代时的开始位置，默认值为 0；end 为迭代时的结束位置，默认值是最后一个元素；step 为迭代步长，默认值为 1；varStatus 代表迭代变量的状态，包括 count（迭代的次数）、index（当前迭代的索引，第一个索引为 0）、first（是否是第一个迭代对象）和 last（是否是最后一个迭代对象）。例如：

```
<table border=1>
            <tr>
                <th>Value</th>
                <th>Square</th>
                <th>Index</th>
            </tr>
            <c:forEach var="x" varStatus="status" begin="0" end="10"
step="2">
                <tr>
                    <td>${x}</td>
                    <td>${x * x}</td>
                    <td>${status.index}</td>
                </tr>
            </c:forEach>
```

```
</table>
```

上述程序运行结果如图 9.3 所示。

http://localhost:8080/ch9/cforEach.jsp

Value	Square	Index
0	0	0
2	4	2
4	16	4
6	36	6
8	64	8
10	100	10

图 9.3 <c:forEach>标签

2．<c:forTokens>标签

<c:forTokens>用于迭代字符串中由分隔符分隔的各成员，它是通过 java.util. StringTokenizer 实例来完成字符串的分隔的，属性 items 和 delims 作为构造 StringTokenizer 实例的参数。语法格式如下：

```
<c:forTokens var="变量名" items="要迭代的 String 对象" delims="指定分隔字符串的分隔符">
            循环体
</c:forTokens>
```

例如：

```
<c:forTokens items="chenheng:lououjun:gongqingzhi" delims=":" var= "name">
      ${name}<br>
</c:forTokens>
```

上述程序运行结果如图 9.4 所示。

图 9.4 <c:forTokens>标签

<c:forTokens>标签与<c:forEach>标签一样，也有 begin、end、step 和 varStatus 属性，并且用法一样，这里就不再赘述了。

9.2.5 函数标签库

在 JSP 页面中，调用 JSTL 中的函数时，需要使用 EL 表达式，调用语法格式如下：

```
${fn:函数名(参数 1，参数 2，…)}
```

下面介绍几个常用的函数。

1．contains 函数

该函数功能是判断一个字符串中是否包含指定的子字符串。如果包含，则返回 True，否则返回 False。其定义如下：

```
contains(string, substring)
```

该函数调用示例代码如下：

```
${fn:contains("I am studying", "am") }
```

上述 EL 表达式将返回 True。

2．containsIgnoreCase 函数

该函数与 contains 函数功能相似，但判断是不区分大小写的。其定义如下：

```
containsIgnoreCase(string, substring)
```

该函数调用示例代码如下：

```
${fn:containsIgnoreCase("I AM studying", "am") }
```

上述 EL 表达式将返回 True。

3．endsWith 函数

该函数功能是判断一个字符串是否以指定的后缀结尾。其定义如下：

```
endsWith(string, suffix)
```

该函数调用示例代码如下：

```
${fn:endsWith("I AM studying", "am") }
```

上述 EL 表达式将返回 False。

4．indexOf 函数

该函数功能是返回指定子字符串在某个字符串中第一次出现时的索引，找不到时，将返回-1。其定义如下：

```
indexOf(string, substring)
```

该函数调用示例代码如下：

```
${fn:indexOf("I am studying", "am") }
```

上述 EL 表达式将返回 2。

5. join 函数

该函数功能是将一个 String 数组中的所有元素合并成一个字符串，并用指定的分隔符分开。其定义如下：

```
join(array, separator)
```

例如，假设一个 String 数组 my，它有三个元素："I""am" 和 "studying"，那么，下列 EL 表达式：

```
${fn:join(my, ",") }
```

将返回 "I,am,studying"。

6. length 函数

该函数功能是返回集合中元素的个数，或者字符串中的字符个数。其定义如下：

```
length(input)
```

该函数调用示例代码如下：

```
${fn:length("aaa")}
```

上述 EL 表达式将返回 3。

7. replace 函数

该函数功能是将字符串中出现的所有 beforestring 用 afterstring 替换，并返回替换后的结果。其定义如下：

```
replace(string, beforestring, afterstring)
```

该函数调用示例代码如下：

```
${fn:replace("I am am studying", "am", "do") }
```

上述 EL 表达式将返回 "I do do studying"。

8. split 函数

该函数功能是将一个字符串，使用指定的分隔符 separator 分离成一个子字符串数组。其定义如下：

```
split(string, separator)
```

该函数调用示例代码如下：

```
<c:set var="my" value="${fn:split('I am studying', ' ') }"/>
<c:forEach var="myArrayElement" items="${my }">
```

```
        ${myArrayElement}<br>
</c:forEach>
```

上述示例代码显示结果如图 9.5 所示。

http://localhost:8080/ch9/testFunctionsjsp.jsp
I
am
studying

图 9.5　split 示例结果

9．startsWith 函数

该函数功能是判断一个字符串是否以指定的前缀开头。其定义如下：

```
startsWith(string, prefix)
```

该函数调用示例代码如下：

```
${fn:startsWith("I AM studying", "am") }
```

上述 EL 表达式将返回 False。

10．substring 函数

该函数功能是返回一个字符串的子字符串。其定义如下：

```
substring(string, begin, end)
```

该函数调用示例代码如下：

```
${fn:substring("abcdef", 1, 3)}
```

上述 EL 表达式将返回“bc”。

11．toLowerCase 函数

该函数功能是将一个字符串转换成它的小写版本。其定义如下：

```
toLowerCase(string)
```

该函数调用示例代码如下：

```
${fn:toLowerCase("I AM studying") }
```

上述 EL 表达式将返回“i am studying”。

12．toUpperCase 函数

该函数功能与 toLowerCase 函数的功能相反，这里不再赘述。

13．trim 函数

该函数功能是将一个字符串开头和结尾的空白去掉。其定义如下：

```
trim(string)
```

该函数调用示例代码如下：

```
${fn:trim("    I AM studying      ") }
```

上述 EL 表达式将返回“I AM studying”。

9.2.6 实践环节

编写一个 JSP 页面，在该页面中使用<c:forEach>标签输出九九乘法表。页面运行效果如图 9.6 所示。

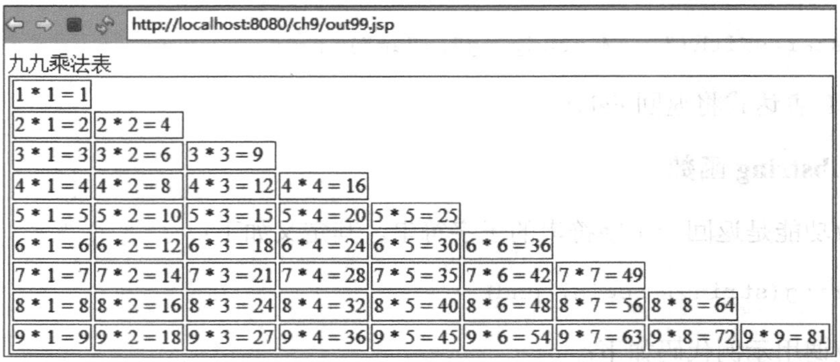

http://localhost:8080/ch9/out99.jsp

九九乘法表

1 * 1 = 1								
2 * 1 = 2	2 * 2 = 4							
3 * 1 = 3	3 * 2 = 6	3 * 3 = 9						
4 * 1 = 4	4 * 2 = 8	4 * 3 = 12	4 * 4 = 16					
5 * 1 = 5	5 * 2 = 10	5 * 3 = 15	5 * 4 = 20	5 * 5 = 25				
6 * 1 = 6	6 * 2 = 12	6 * 3 = 18	6 * 4 = 24	6 * 5 = 30	6 * 6 = 36			
7 * 1 = 7	7 * 2 = 14	7 * 3 = 21	7 * 4 = 28	7 * 5 = 35	7 * 6 = 42	7 * 7 = 49		
8 * 1 = 8	8 * 2 = 16	8 * 3 = 24	8 * 4 = 32	8 * 5 = 40	8 * 6 = 48	8 * 7 = 56	8 * 8 = 64	
9 * 1 = 9	9 * 2 = 18	9 * 3 = 27	9 * 4 = 36	9 * 5 = 45	9 * 6 = 54	9 * 7 = 63	9 * 8 = 72	9 * 9 = 81

图 9.6　使用<c:forEach>打印九九乘法表

9.3 本章小结

本章重点介绍了 EL 表达式、JSTL 核心标签库以及 JSTL 函数标签库的用法。 EL 与 JSTL 的应用大大提高了编程效率，并且降低了维护难度。

习 题 9

1．在 Web 应用程序中有以下程序代码段，执行后转发到某个 JSP 页面：

```
ArrayList<String>  dogNames = new ArrayList<String>();
dogNames.add("goodDog");
request.setAttribute("dogs", dogNames);
```

以下（　　）选项可以正确地使用 EL 取得数组中的值。

A. ${ dogs .0}
B. ${ dogs [0]}
C. ${ dogs .[0]}
D. ${ dogs "0"}

2. ( ) JSTL 标签可以实现 Java 程序中的 if 语句功能。
A. <c:set>
B. <c:out>
C. <c:forEach>
D. <c:if>

3. ( ) 不是 EL 的隐含对象。
A. request
B. pageScope
C. sessionScope
D. applicationScope

4. ( ) JSTL 标签可以实现 Java 程序中的 for 语句功能。
A. <c:set>
B. <c:out>
C. <c:forEach>
D. <c:if>

第 10 章 名片管理系统的设计与实现

chapter 10

本章主要内容

- 系统设计
- 数据库设计
- 系统管理
- 组件设计
- 系统实现

本章系统使用 Spring MVC 框架实现各个模块，Web 引擎为 Tomcat 8.5，数据库采用的是 MySQL 5.x，集成开发环境为 Eclipse。

10.1 系统设计

10.1.1 系统功能需求

名片管理系统是针对注册用户使用的系统。系统提供的功能如下：

（1）非注册用户可以注册为注册用户。

（2）成功注册的用户，可以登录系统。

（3）成功登录的用户，可以添加、修改、删除以及浏览自己客户的名片信息。

（4）成功登录的用户，可以在个人中心查看自己的基本信息和修改密码。

10.1.2 系统模块划分

用户登录成功后，进入管理主页面（main.jsp），可以对自己的客户名片进行管理。系统模块划分如图 10.1 所示。

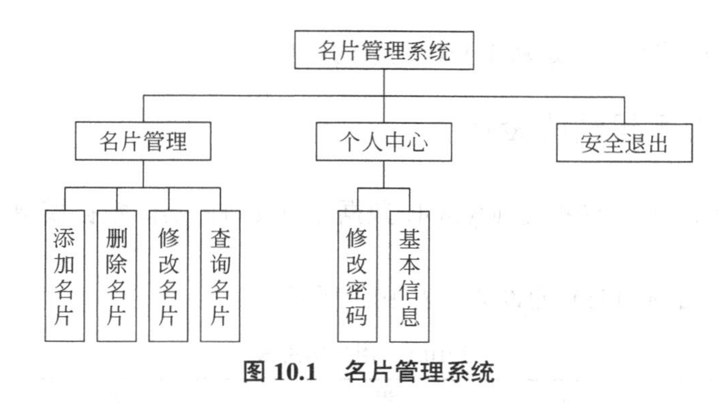

图 10.1　名片管理系统

10.2　数据库设计

系统采用加载纯 Java 数据库驱动程序的方式连接 MySQL 5.5 数据库。在 MySQL 5.x 的数据库 card 中，共创建两张与系统相关的数据表：usertable 和 cardinfo。

10.2.1　数据库概念结构设计

根据系统设计与分析，可以设计出如下数据结构。

1．用户

包括用户名和密码，注册用户名唯一。

2．名片

包括 ID、名称、电话、邮箱、单位、职务、地址、Logo 以及所属用户。其中，ID 唯一，“所属用户”与“1．用户”中的用户名关联。

根据以上数据结构，结合数据库设计特点，可画出如图 10.2 所示的数据库概念结构图。

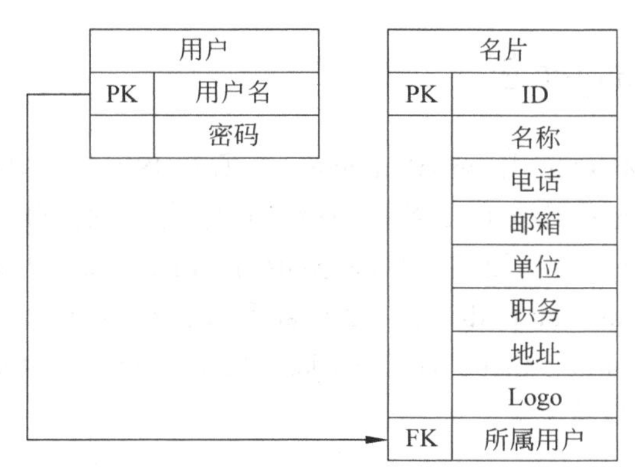

图 10.2　数据库概念结构图

其中，ID 为正整数，值是从 1 开始递增的序列。

10.2.2 数据库逻辑结构设计

将数据库概念结构图转换为 MySQL 数据库所支持的实际数据模型，即数据库的逻辑结构。

用户信息表（usertable）的设计如表 10.1 所示。

表 10.1 用户信息表

字 段	含 义	类 型	长 度	是否为空
userName	用户名（PK）	varchar	50	no
password	密码	varchar	20	no

名片信息表（cardinfo）的设计如表 10.2 所示。

表 10.2 名片信息表

字 段	含 义	类 型	长 度	是否为空
id	编号（PK）	int	11	no
name	名称	varchar	50	no
telephone	电话	varchar	20	no
email	邮箱	varchar	20	
company	单位	varchar	50	
post	职务	varchar	50	
address	地址	varchar	50	
logo	图片	varchar	25	
userName	所属用户	varchar	50	no

10.3 系 统 管 理

10.3.1 导入相关的 jar 包

新建一个 Spring MVC 应用 cardsManage，在所有 JSP 页面中尽量使用 EL 表达式和 JSTL 标签。又因为系统采用纯 Java 数据库驱动程序连接 MySQL 5.x，所以除了 Spring MVC 框架所需 jar 包外，还需要将数据库驱动程序 mysql-connector-java-5.1.26-bin.jar、JSTL 对应的 jar 包 taglibs-standard-impl-1.2.5.jar 和 taglibs-standard-spec-1.2.5.jar，以及文件上传需要的 jar 包 commons-fileupload-1.3.1.jar 和 commons-io-2.4.jar 复制到 cardsManage/WebRoot/WEB-INF/lib 目录下。

10.3.2　JSP 页面管理

由于篇幅受限，本章仅附上 JSP 和 Java 文件的代码。

1．管理主页面

注册用户在浏览器地址栏中输入"http://localhost:8080/cardsManage/"访问系统首页，然后进入登录页面。登录成功后，进入管理主页面（main.jsp）。main.jsp 的运行效果如图 10.3 所示。

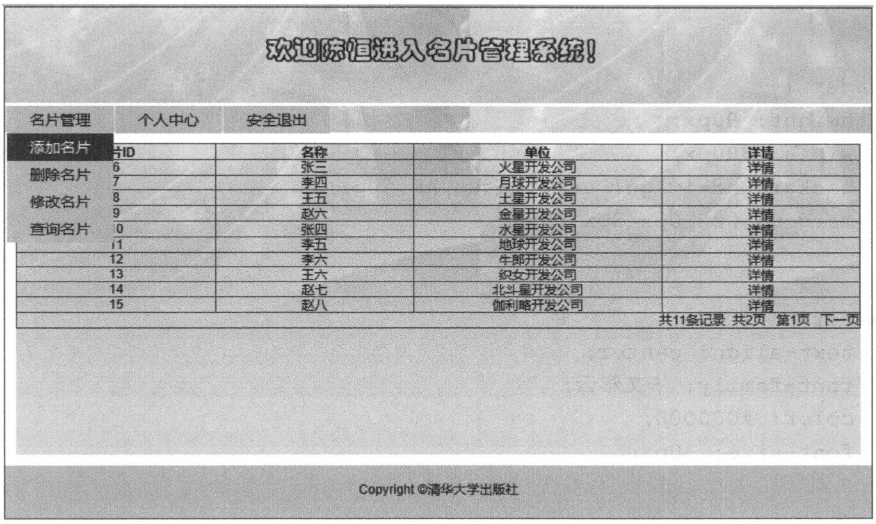

图 10.3　管理主页面

管理主页面 main.jsp 的代码如下：

```
<%@ page language="java" contentType="text/html; charset=UTF-8"
pageEncoding="UTF-8"%>
<%
String path = request.getContextPath();
String basePath = request.getScheme()+"://"+request.getServerName()+":"+
request.getServerPort()+path+"/";
%>
<!DOCTYPE html PUBLIC "-//W3C//DTD HTML 4.01 Transitional//EN"
"http://www.w3.org/TR/html4/loose.dtd">
<html>
<head>
<base href="<%=basePath%>">
<title>后台主页面</title>
<style type="text/css">
* {
```

```
    margin: 0px;
    padding: 0px;
}

body {
    font-family: Arial, Helvetica, sans-serif;
    font-size: 12px;
    margin: 0px auto;
    height: auto;
    width: 800px;
    border: 1px solid #006633;
}

#header {
    height: 90px;
    width: 800px;
    background-image: url(images/bb.jpg);
    margin: 0px 0px 3px 0px;
}

#header h1 {
    text-align: center;
    font-family: 华文彩云;
    color: #000000;
    font-size: 30px;
}

#navigator {
    height: 25px;
    width: 800px;
    font-size: 14px;
    background-image: url(images/bb.jpg);
}
#navigator ul {
    list-style-type: none;
}
#navigator li {
    float: left;
    position: relative;
}
#navigator li a {
    color: #000000;
    text-decoration: none;
    padding-top: 4px;
    display: block;
    width: 98px;
```

```
    height: 22px;
    text-align: center;
    background-color: PaleGreen;
    margin-left: 2px;
}
#navigator li a:hover {
    background-color: #006633;
    color: #FFFFFF;
}
#navigator ul li ul {
  visibility: hidden;
  position: absolute;
}

#navigator ul li:hover ul,
#navigator ul a:hover ul{
  visibility: visible;
}

#content {
    height: auto;
    width: 780px;
    padding: 10px;
}

#content iframe {
    height: 300px;
    width: 780px;
}

#footer {
    height: 30px;
    width: 780px;
    line-height: 2em;
    text-align: center;
    background-color: PaleGreen;
    padding: 10px;
}
</style>
</head>
<body>
    <div id="header">
        <br>
        <br>
        <h1>欢迎${sessionScope.user.username}进入名片管理系统！</h1>
    </div>
```

```
    <div id="navigator">
        <ul>
            <li><a>名片管理</a>
                <ul>
                    <li><a href="card/input" target="center">添加名片
</a></li>
                <li><a href="card/query?act=deleteSelect" target="center">
删除名片</a></li>
                <li><a href="card/query?act=updateSelect" target="center">
修改名片</a></li>
                    <li><a href="card/query" target="center">查询名片</a>
</li>
                </ul>
            </li>
            <li><a>个人中心</a>
                <ul>
                    <li><a href="user/updatePWD" target="_top">修改密码
</a></li>
                    <li><a href="user/userInfo" target="center">基本信息
</a></li>
                </ul>
            </li>
            <li><a href="user/exit">安全退出</a></li>
        </ul>

    </div>
    <div id="content">
        <iframe src="card/query" name="center" frameborder="0"> </iframe>
    </div>
    <div id="footer">Copyright ©清华大学出版社</div>
</body>
</html>
```

2．未知错误页面

当Java程序运行出现未知异常时，系统将统一异常处理，然后跳转到对应的error.jsp，具体代码如下：

```
<%@ page language="java" contentType="text/html; charset=UTF-8"
pageEncoding="UTF-8" isErrorPage="true"%>
<%
String path = request.getContextPath();
String basePath = request.getScheme()+"://"+request.getServerName()+":"+
request.getServerPort()+path+"/";
%>
<!DOCTYPE html PUBLIC "-//W3C//DTD HTML 4.01 Transitional//EN"
```

```
"http://www.w3.org/TR/html4/loose.dtd">
<html>
<head>
<base href="<%=basePath%>">
<meta http-equiv="Content-Type" content="text/html; charset=UTF-8">
<title>Insert title here</title>
</head>
<body>
<H1>未知错误：</H1><%=exception%>
<H2>错误内容：</H2>
<%
    exception.printStackTrace(response.getWriter());
%>
</body>
</html>
```

3. 系统首页面

index.jsp 的具体代码如下：

```
<%@ page language="java" contentType="text/html; charset=UTF-8"
pageEncoding="UTF-8"%>
<%
String path = request.getContextPath();
String basePath = request.getScheme()+"://"+request.getServerName()+":"+
request.getServerPort()+path+"/";
%>
<!DOCTYPE html PUBLIC "-//W3C//DTD HTML 4.01 Transitional//EN"
"http://www.w3.org/TR/html4/loose.dtd">
<html>
  <head>
    <base href="<%=basePath%>">
    <meta http-equiv="Content-Type" content="text/html; charset=UTF-8">
    <title>My JSP 'index.jsp' starting page</title>
  </head>
  <body>
    没注册的用户，请<a href="register">注册</a>！<br>
    已注册的用户，去<a href="login">登录</a>！
  </body>
</html>
```

在首页 index.jsp 中有两个超链接，处理超链接请求的 Controller 是 IndexController，具体代码如下：

```
package com.controller;
import org.springframework.stereotype.Controller;
```

```
import org.springframework.ui.Model;
import org.springframework.web.bind.annotation.RequestMapping;
import com.domain.User;
@Controller
@RequestMapping("/")
public class IndexController {
    @RequestMapping("/login")
    public String login(Model model) {
        model.addAttribute("user", new User());
        //跳转到 user 目录下
        return "/user/login";
    }
    @RequestMapping("/register")
    public String register(Model model) {
        model.addAttribute("user", new User());
        return "/user/register";
    }
}
```

10.3.3 包管理

本系统的包层次结构如图 10.4 所示。

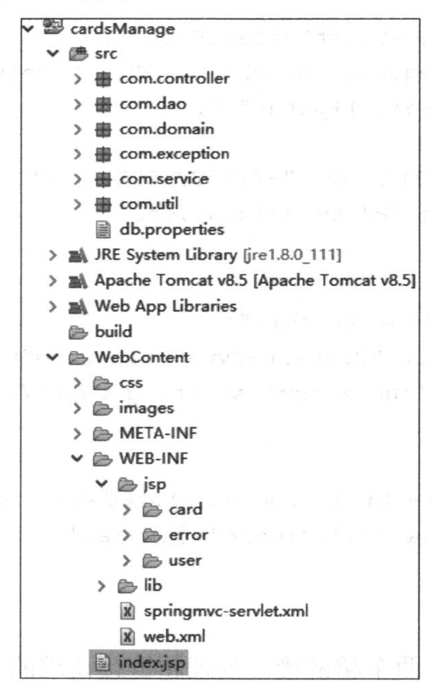

图 10.4 包层次结构图

1．controller 包

该包存放系统中所有控制器类，包括名片管理的控制器类和个人中心的控制器类。

2．dao 包

dao 包中存放的 Java 程序用于实现数据库的操作。其中，BaseDao 是一个父类，该类负责连接数据库、统一操作数据库；CardDao 是 BaseDao 的一个子类，有关名片管理的数据访问在该类中；UserDao 是 BaseDao 的另一个子类，有关用户的数据访问在该类中。

3．domain 包

该包中有两个领域模型类：Card 封装名片信息，User 封装用户信息。

4．exception 包

该包有两个异常处理类：LoginNoException 和 MyExceptionHandler。LoginNoException 类是未登录异常类，MyExceptionHandler 是统一异常处理类。

5．service 包

该包负责业务处理，是控制器和 dao 的桥梁，其中有 Service 接口和 Service 实现类。

6．util 包

该包中 MyUtil 类是获得一个时间字符串的工具类。

10.3.4 配置文件管理

1．web.xml

web.xml 除了部署 DispatcherServlet 之外，还部署了字符编码过滤器（防止中文乱码）CharacterEncodingFilter。

web.xml 文件的代码如下：

```
<?xml version="1.0" encoding="UTF-8" ?>
<web-app
xmlns:xsi="http://www.w3.org/2001/XMLSchema-instance"
xmlns="http://xmlns.jcp.org/xml/ns/javaee"
xsi:schemaLocation="http://xmlns.jcp.org/xml/ns/javaee
http://xmlns.jcp.org/xml/ns/javaee/web-app_3_1.xsd"
id="WebApp_ID" version="3.1">
  <display-name>cardManage</display-name>
  <welcome-file-list>
    <welcome-file>index.html</welcome-file>
```

```
    <welcome-file>index.htm</welcome-file>
    <welcome-file>index.jsp</welcome-file>
    <welcome-file>default.html</welcome-file>
    <welcome-file>default.htm</welcome-file>
    <welcome-file>default.jsp</welcome-file>
  </welcome-file-list>

<!--配置 springmvcDispatcherServlet-->
<servlet>
    <servlet-name>springmvc</servlet-name>
    <servlet-class>org.springframework.web.servlet.DispatcherServlet
</servlet-class>
    <load-on-startup>1</load-on-startup>
</servlet>
<servlet-mapping>
    <servlet-name>springmvc</servlet-name>
    <url-pattern>/</url-pattern>
</servlet-mapping>

 <!-- 避免中文乱码 -->
  <filter>
        <filter-name>characterEncodingFilter</filter-name>
        <filter-class>org.springframework.web.filter.CharacterEncodingFilter
</filter-class>
        <init-param>
            <param-name>encoding</param-name>
            <param-value>UTF-8</param-value>
        </init-param>
        <init-param>
            <param-name>forceEncoding</param-name>
            <param-value>true</param-value>
        </init-param>
    </filter>
    <filter-mapping>
        <filter-name>characterEncodingFilter</filter-name>
        <url-pattern>/*</url-pattern>
    </filter-mapping>
</web-app>
```

2．springmvc-servlet.xml

该配置文件托管了Spring MVC的相关bean，包括视图解析器internalResourceViewResolver、数据源dataSource、统一异常处理MyExceptionHandler以及上传文件分解器multipartResolver，具体配置代码如下：

```
<?xml version="1.0" encoding="UTF-8" ?>
<beans xmlns="http://www.springframework.org/schema/beans"
    xmlns:xsi="http://www.w3.org/2001/XMLSchema-instance"
    xmlns:p="http://www.springframework.org/schema/p"
    xmlns:context="http://www.springframework.org/schema/context"
    xmlns:mvc="http://www.springframework.org/schema/mvc"
    xmlns:jdbc="http://www.springframework.org/schema/jdbc"
    xsi:schemaLocation="
     http://www.springframework.org/schema/beans
     http://www.springframework.org/schema/beans/spring-beans.xsd
        http://www.springframework.org/schema/context
        http://www.springframework.org/schema/context/spring-context.xsd
        http://www.springframework.org/schema/mvc
        http://www.springframework.org/schema/mvc/spring-mvc.xsd
        http://www.springframework.org/schema/jdbc
        http://www.springframework.org/schema/jdbc/spring-jdbc.xsd">
      <!-- 附加 props 配置文件 -->
     <context:property-placeholder location="classpath:db.properties" />

    <!-- scan the package and the sub package -->
    <context:component-scan base-package="com.controller"/>
    <context:component-scan base-package="com.service"/>
    <context:component-scan base-package="com.dao"/>

    <mvc:annotation-driven />
    <!-- 静态资源需要单独处理，不需要 dispatcher servlet -->
    <mvc:resources location="/css/" mapping="/css/**"></mvc:resources>
    <mvc:resources location="/images/" mapping="/images/**"></mvc:resources>
    <!-- 查看名片图片时，logos 文件夹不需要 dispatcher servlet -->
    <mvc:resources location="/logos/" mapping="/logos/**"> </mvc:resources>
    <!-- 配置视图解析器 -->
    <bean class="org.springframework.web.servlet.view.InternalResource_
ViewResolver"
            id="internalResourceViewResolver">
        <!-- 前缀 -->
        <property name="prefix" value="/WEB-INF/jsp" />
        <!-- 后缀 -->
        <property name="suffix" value=".jsp" />
    </bean>

     <bean id="dataSource" class="org.apache.tomcat.dbcp.dbcp2.Basic_
DataSource">
         <property name="driverClassName" value="${db.driverClassName}" />
         <property name="url" value="${db.url}" />
         <property name="username" value="${db.username}" />
         <property name="password" value="${db.password}" />
```

```
        <!-- 读取classpath:db.properties属性文件 -->
    </bean>

   <bean id="jdbcTemplate" class="org.springframework.jdbc.core.Jdbc_
Template">
        <property name="dataSource" ref="dataSource"/>
    </bean>
    <!-- 托管MyExceptionHandler -->
    <bean class="com.exception.MyExceptionHandler"/>

    <!--使用spring的CommosMultipartResolver配置MultipartResolver用于文
件上传 -->
     <bean id="multipartResolver" class="org.springframework.web.multipart.
commons.CommonsMultipartResolver"
        p:defaultEncoding="UTF-8"
        p:maxUploadSize="5400000"
        p:uploadTempDir="fileUpload/temp"
    >
     <!--D:\spring mvc workspace\.metadata\.plugins\org.eclipse.wst.server.
core\tmp0\wtpwebapps\fileUpload -->
   </bean>
   <!-- defaultEncoding="UTF-8" 是请求的编码格式，默认为iso-8859-1
    maxUploadSize="5400000" 是允许上传文件的最大值，单位为字节
    uploadTempDir="fileUpload/temp" 为上传文件的临时路径 -->
</beans>
```

10.4 组件设计

本系统的组件包括工具类、异常处理类、登录权限控制器和数据库统一操作（BaseDao）。

10.4.1 工具类

工具类MyUtil的代码如下：

```
package com.util;
import java.text.SimpleDateFormat;
import java.util.Date;
public class MyUtil {
    /**
     * 获得一个以时间字符串为标准的ID，固定长度是17位
     */
    public static String getStringID(){
        String id=null;
```

```
            Date date=new Date();
            SimpleDateFormat sdf=new SimpleDateFormat("yyyyMMddHHmmssSSS");
            id=sdf.format(date);
            return id;
        }
    }
```

10.4.2 统一异常处理

未登录异常类 LoginNoException 的代码如下：

```
package com.exception;
public class LoginNoException extends Exception{
    private static final long serialVersionUID = 1L;
    public LoginNoException(String message){
        super(message);
    }
}
```

统一异常处理类 MyExceptionHandler 的代码如下：

```
package com.exception;
import java.sql.SQLException;
import java.util.HashMap;
import java.util.Map;
import javax.servlet.http.HttpServletRequest;
import javax.servlet.http.HttpServletResponse;
import org.springframework.web.servlet.HandlerExceptionResolver;
import org.springframework.web.servlet.ModelAndView;
import com.domain.User;
public class MyExceptionHandler implements HandlerExceptionResolver {
    @Override
    public  ModelAndView  resolveException(HttpServletRequest  arg0,
HttpServletResponse arg1, Object arg2,Exception arg3) {
        Map<String, Object> model = new HashMap<String, Object>();
        model.put("ex", arg3);
        // 根据不同错误转向不同页面
        if(arg3 instanceof SQLException) {
         return new ModelAndView("/error/sql-error", model);
        }else if(arg3 instanceof LoginNoException){
         //登录页面需要 user 对象
         arg0.setAttribute("user", new User());
         return new ModelAndView("/user/login", model);
        }else{
         return new ModelAndView("/error/error", model);
        }
```

```
    }
}
```

10.4.3 登录权限控制器

本系统设计了一个BaseController，在该基类中使用@ModelAttribute声明了一个非处理方法，在该方法中判定用户是否已登录。需要判断用户登录权限的控制器类，继承BaseController即可。BaseController类的代码如下：

```
package com.controller;
import javax.servlet.http.HttpServletRequest;
import javax.servlet.http.HttpSession;
import org.springframework.stereotype.Controller;
import org.springframework.web.bind.annotation.ModelAttribute;
import com.exception.LoginNoException;
@Controller
public class BaseController {
    /**
     * 登录权限控制，处理方法执行前执行该方法
     * @throws LoginNoException
     */
    @ModelAttribute
   public void isLogin(HttpSession session, HttpServletRequest request)
throws LoginNoException {
      if(session.getAttribute("user") == null){
          throw new LoginNoException("没有登录");
      }
    }
}
```

10.4.4 数据库统一操作

本系统有关数据库操作的Java类位于包dao中。为了方便管理，系统中定义了一个数据库统一操作类BaseDao，其他数据库访问类继承该类即可。BaseDao.java的代码如下：

```
package com.dao;
import java.sql.ResultSet;
import java.sql.ResultSetMetaData;
import java.sql.SQLException;
import java.util.HashMap;
import java.util.List;
import java.util.Map;
import org.springframework.beans.factory.annotation.Autowired;
```

```
import org.springframework.jdbc.core.JdbcTemplate;
import org.springframework.jdbc.core.RowMapper;
public class BaseDao {
    @Autowired
    private JdbcTemplate jdbcTemplate;
    /**
     * @description 添加、修改、删除操作
     * @param String sql 代表 SQL 文
     * @param 根据参数进行增删改，注意参数 Object[] arg0 元素的个数和顺序一定与
String sql 中的通配符？一一对应，如果 SQL 语句中没有通配符请传递 null
     * @return boolean 更新成功返回 true，否则 false
     */
    public boolean updateByParam(String sql, Object[] arg0) {
        //SQL 语句中没有通配符
        if(arg0 == null || arg0.length == 0){
            if(jdbcTemplate.update(sql)>0)
                return true;
            else
                return false;
        }
        if(jdbcTemplate.update(sql,arg0)>0)
            return true;
        else
            return false;
    }

    /**
     * @description support select SQL Statement by param
     * @param String sql 代表 SQL 文
     * @param 根据参数进行查询，注意参数 Object[] arg0 元素的个数和顺序一定与
String sql 中的通配符？一一对应，如果 SQL 语句中没有通配符请传递 null
     * @return List<Map<String, Object>>: 查询结果
     */
    @SuppressWarnings("unchecked")
    public List<Map<String, Object>> findByParam(String sql,Object arg0[]){
        if(arg0 == null || arg0.length == 0){
            return jdbcTemplate.query(sql, new IRowMapper());
        }
        return jdbcTemplate.query(sql, new IRowMapper(),arg0);
    }
    /**
     * 数据库表的列名与 map 的 key 相同
     */
    @SuppressWarnings("rawtypes")
    public class IRowMapper implements RowMapper{
        @Override
```

```
        public Map<String, Object> mapRow(ResultSet rs, int rowNum) throws
SQLException {
            Map<String, Object> row = new HashMap<String, Object>();
            ResultSetMetaData rsMetaData = rs.getMetaData();
            for(int i=1,size=rsMetaData.getColumnCount(); i<=size; i++){
                //查询中，数据表的字段名和页面里面要一模一样
                row.put(rsMetaData.getColumnLabel(i).toLowerCase(),
rs.getObject(i));
            }
            return row;
        }
    }
}
```

10.5 名片管理

与系统相关的 CSS 和图片位于 WebContent 目录下，与名片管理相关的 JSP 页面位于 WEB-INF/jsp/card 目录下。

10.5.1 Controller 实现

CardController 类负责处理“名片管理”的功能，包括添加、修改、删除、查询等。具体代码如下：

```
package com.controller;
import java.io.File;
import java.util.List;
import java.util.Map;
import javax.servlet.http.HttpServletRequest;
import javax.servlet.http.HttpSession;
import org.springframework.beans.factory.annotation.Autowired;
import org.springframework.stereotype.Controller;
import org.springframework.ui.Model;
import org.springframework.web.bind.annotation.ModelAttribute;
import org.springframework.web.bind.annotation.RequestMapping;
import com.domain.Card;
import com.domain.User;
import com.service.CardsService;
import com.util.MyUtil;
@Controller
@RequestMapping("/card")
public class CardController extends BaseController{
    //依赖注入 Service 进行后台处理
    @Autowired
```

```
    private CardsService cardsService;
    /**
     * add 页面初始化
     */
    @RequestMapping("/input")
    public String addInput(Model model){
        model.addAttribute("card", new Card());
        return "/card/addCard";
    }
    /**
     * 查询名片，包括修改、删除的查询
     */
    @RequestMapping("/query")
    public String query(HttpSession session, Model model, Integer pageCur,
String act){
        List<Map<String, Object>> allCards = cardsService.query
(getUserName(session));
        int temp = allCards.size();
        model.addAttribute("totalCount", temp);
        int totalPage = 0;
        if (temp == 0) {
            totalPage = 0;//总页数
        } else {
            //返回大于或者等于指定表达式的最小整数
            totalPage = (int) Math.ceil((double) temp / 10);
        }
        if (pageCur == null) {
            pageCur = 1;
        }
        if ((pageCur - 1) * 10 > temp) {//第一页和最后一页时，使用。比如第一
                                       //页点击上一页，最后一页点击下一页。
            pageCur = pageCur - 1;
        }

        //分页查询
        allCards = cardsService.queryByPage(pageCur, getUserName(session));

        model.addAttribute("allCards", allCards);
        model.addAttribute("totalPage", totalPage);
        model.addAttribute("pageCur", pageCur);
        //删除查询
        if("deleteSelect".equals(act)){
            return "/card/deleteSelect";
        }
        //修改查询
        else if("updateSelect".equals(act)){
```

```
            return "/card/updateSelect";
        }else{
            return "/card/queryCards";
        }
    }
    /**
     * 添加与修改页面
     */
    @RequestMapping("/add")
    public String add(@ModelAttribute Card card, HttpServletRequest
request, HttpSession session, String updateAct){
        /*上传文件的保存位置"/logos", 该位置是指
    workspace\.metadata\.plugins\org.eclipse.wst.server.core\tmp0\wtpw
ebapps,
        发布后使用*/
        //防止文件名重名
        String newFileName = "";
        String fileName = card.getLogo().getOriginalFilename();
        //选择了文件
        if(fileName.length() > 0){
            String realpath = request.getServletContext().getRealPath
("logos");
            //实现文件上传
            String fileType=fileName.substring(fileName.lastIndexOf('.'));
            //防止文件名重名
            newFileName = MyUtil.getStringID() + fileType;
            File targetFile = new File(realpath, newFileName);
            if(!targetFile.exists()){
              targetFile.mkdirs();
           }
            //上传
           try {
            card.getLogo().transferTo(targetFile);
           } catch (Exception e) {
              e.printStackTrace();
           }
        }
        //修改
        if("update".equals(updateAct)){//updateAct不能与act重名,因为使用了转发
            //修改到数据库
           if(cardsService.update(card, newFileName)){
            return "forward:/card/query?act=updateSelect";
           }else{
            return "/card/updateCard";
           }
```

```
    }else{//添加
        //保存到数据库
        if(cardsService.add(card, newFileName, getUserName(session))){
            //转发到查询的 controller
            return "forward:/card/query";
        }else{
            return "/card/addCard";
        }
    }
}
/**
 * 查询一个名片
 */
@RequestMapping("selectACard")
public String selectACard(Model model, String id, String act){
    Map<String, Object> mp = cardsService.selectACard(id);
    model.addAttribute("acard", mp);
    //修改页面
    if("updateAcard".equals(act)){
        Card card = new Card();
        card.setId((Integer)mp.get("id"));
        card.setName((String)mp.get("name"));
        card.setAddress((String)mp.get("address"));
        card.setCompany((String)mp.get("company"));
        card.setEmail((String)mp.get("email"));
        //从数据库取出的文件名
        card.setFileName((String)mp.get("logo"));
        card.setPost((String)mp.get("post"));
        card.setTelephone((String)mp.get("telephone"));
        model.addAttribute("card", card);
        return "/card/updateCard";
    }
    //详情页面
    return "/card/detail";
}
/**
 * 删除一个名片
 */
@RequestMapping("deleteACard")
public String deleteACard(String id){
    cardsService.deleteAcard(id);
    return "forward:/card/query?act=deleteSelect";
}
/**
 * 删除多个名片
```

```
     */
    @RequestMapping("deleteCards")
    public String deleteCards(String[] ids){
        cardsService.deleteCards(ids);
        return "forward:/card/query?act=deleteSelect";
    }
    /**
     * 获得登录用户的用户名(非处理请求方法)
     */
    public String getUserName(HttpSession session){
         User user = (User)session.getAttribute("user");
         return user.getUsername();
    }
}
```

10.5.2 Service 实现

与名片管理相关的 Service 接口和实现类分别为 CardsService 和 CardsServiceImpl。控制器获取一个请求后，需要调用 Service 层中业务处理方法，在 Service 层中需要访问层 dao。所以，Service 层是控制器层和 dao 层的桥梁。

CardsService 接口代码如下：

```
package com.service;
import java.util.List;
import java.util.Map;
import com.domain.Card;
public interface CardsService {
    List<Map<String, Object>> query(String userName);
    List<Map<String, Object>>  queryByPage(int pageCur, String userName);
    boolean add(Card card, String newFileName, String userName);
    boolean update(Card card, String newFileName);
    Map<String, Object> selectACard(String id);
    boolean deleteAcard(String id);
    boolean deleteCards(String[] id);
}
```

CardsServiceImpl 实现类的代码如下：

```
package com.service;
import java.util.List;
import java.util.Map;
import org.springframework.beans.factory.annotation.Autowired;
import org.springframework.stereotype.Service;
import com.dao.CardDao;
import com.domain.Card;
```

```
//注册为一个服务
@Service
public class CardsServiceImpl implements CardsService{
    @Autowired
    private CardDao cardDao;
    /**
     * 查询用户下所有名片信息
     */
    @Override
    public List<Map<String, Object>> query(String userName) {
        return cardDao.query(userName);
    }
    /**
     * 分页查询用户下名片信息
     */
    @Override
    public List<Map<String, Object>>  queryByPage(int pageCur, String
userName){
        return cardDao.queryByPage(pageCur, userName);
    }
    /**
     *给某用户添加他的名片
     */
    @Override
    public boolean add(Card card, String newFileName, String userName) {
        return cardDao.add(card, newFileName, userName);
    }
    /**
     * 查询一个名片
     */
    @Override
    public Map<String, Object> selectACard(String id) {
        return cardDao.selectACard(id);
    }
    /**
     * 删除一个名片
     */
    @Override
    public boolean deleteAcard(String id) {
        return cardDao.deleteAcard(id);
    }
    /**
     * 删除多个名片
     */
    @Override
    public boolean deleteCards(String[] id) {
```

```
            return cardDao.deleteCards(id);
        }
        /**
         *修改名片
         */
        @Override
        public boolean update(Card card, String newFileName) {
            return cardDao.update(card, newFileName);
        }
    }
```

10.5.3 Dao 实现

Dao 层是数据访问层，与名片管理相关的数据访问层为 CardDao，具体代码如下：

```
package com.dao;
import java.util.ArrayList;
import java.util.List;
import java.util.Map;
import org.springframework.stereotype.Repository;
import com.domain.Card;
//注解为存储层
@Repository("cardDao")
public class CardDao extends BaseDao{
    /**
     * 查询用户下所有名片信息
     */
    public List<Map<String, Object>> query(String userName) {
        String sql = " select * from cardinfo where userName=? ";
        Object obj[] = {userName};
        return findByParam(sql, obj);
    }
    /**
     * 分页查询用户下名片信息
     */
    public List<Map<String, Object>>  queryByPage(int pageCur, String
userName){
        String sql = " select id,name, "
                + "  telephone,email, company, post, address, logo "
                + " from cardinfo  where userName=? limit ?,? ";
        Object obj[] = {userName, (pageCur-1)*10,10};//10 为每页个数
        return findByParam(sql, obj);
    }
    /**
     * 添加名片信息
```

```
 */
public boolean add(Card card, String newFileName, String userName){
    //其中 ID 为自增
    String sql = "insert into cardinfo values(null,?,?,?,?,?,?,?,?)";
    //数组元素与问号对应
    Object obj[] = {
        card.getName(),
        card.getTelephone(),
        card.getEmail(),
        card.getCompany(),
        card.getPost(),
        card.getAddress(),
        newFileName,
        userName
    };
    return updateByParam(sql, obj);
}
/**
 *修改名片
 */
public boolean update(Card card, String newFileName) {
    String sql = "update cardinfo set "
            + " name=?,"
            + " telephone=?,"
            + " email=?,"
            + " company=?,"
            + " post=?,"
            + " address=? ";
    ArrayList<Object> al = new ArrayList<Object>();
    al.add(card.getName());
    al.add(card.getTelephone());
    al.add(card.getEmail());
    al.add(card.getCompany());
    al.add(card.getPost());
    al.add(card.getAddress());
    //修改了图片
    if(newFileName.length() > 0){
        sql = sql + ", logo=? ";
        al.add(newFileName);
    }
    al.add(card.getId());
    sql = sql + " where id=? ";
    return updateByParam(sql, al.toArray());
}
/**
 * 查询一个名片
```

```
     */
    public Map<String, Object> selectACard(String id) {
        String sql = " select * from cardinfo where id = ? ";
        Object obj[] = {id};
        return findByParam(sql, obj).get(0);
    }
    /**
     * 删除一个名片
     */
    public boolean deleteAcard(String id){
        String sql = " delete from cardinfo where id = ? ";
        Object obj[] = {id};
        return updateByParam(sql, obj);
    }
    /**
     * 删除多个名片
     */
    public boolean deleteCards(String[] id){
        String sql = "delete from cardinfo where id in ( ";
        for (int i = 0; i < id.length - 1; i++) {
            sql = sql + id[i] + ",";
        }
        sql = sql + id[id.length - 1] + " )";
        return updateByParam(sql, null);
    }
}
```

10.5.4 添加名片

用户输入客户名片的姓名、电话、E-Mail、单位、职务、地址、Logo 后，单击“提交”按钮实现添加。如果成功，则跳转到查询视图；如果失败，则回到添加视图。

addCard.jsp 页面实现添加名片信息的输入界面，如图 10.5 所示。

图 10.5 添加名片页面

addCard.jsp 的代码如下：

```
<%@ page language="java" contentType="text/html; charset=UTF-8"
pageEncoding="UTF-8"%>
```

```
<%@ taglib prefix="form" uri="http://www.springframework.org/tags/form" %>
<%
String path = request.getContextPath();
String basePath = request.getScheme() + "://" + request.getServerName()
+ ":" + request.getServerPort() + path + "/";
%>
<!DOCTYPE  html  PUBLIC  "-//W3C//DTD  HTML  4.01  Transitional//EN"
"http://www.w3.org/TR/html4/loose.dtd">
<html>
<head>
<base href="<%=basePath%>">
<title>addCard.jsp</title>
<link href="css/common.css" type="text/css" rel="stylesheet">
</head>
<body>
    <form:form  action="card/add"  method="post"  modelAttribute="card"
enctype="multipart/form-data">
        <table border=1 style="border-collapse: collapse">
            <caption>
                <font size=4 face=华文新魏>添加名片</font>
            </caption>
            <tr>
                <td>姓名<font color="red">*</font></td>
                <td>
                    <form:input path="name"/>
                </td>
            </tr>
            <tr>
                <td>电话<font color="red">*</font></td>
                <td>
                    <form:input path="telephone"/>
                </td>
            </tr>
            <tr>
                <td>E-Mail</td>
                <td>
                    <form:input path="email"/>
                </td>
            </tr>
            <tr>
                <td>单位</td>
                <td>
                    <form:input path="company"/>
                </td>
            </tr>
            <tr>
```

```
                    <td>职务</td>
                    <td>
                        <form:input path="post"/>
                    </td>
                </tr>
                <tr>
                    <td>地址</td>
                    <td>
                        <form:input path="address"/>
                    </td>
                </tr>
                <tr>
                    <td>logo</td>
                    <td>
                        <input type="file" name="logo"/>
                    </td>
                </tr>
                <tr>
                    <td align="center">
                        <input type="submit" value="提交"/>
                    </td>
                    <td align="left">
                        <input type="reset" value="重置"/>
                    </td>
                </tr>
            </table>
        </form:form>
        ${errorMessage }
    </body>
    </html>
```

单击图 10.5 中的"提交"按钮，将添加请求通过"card/add"提交给控制器类 CardController（第 10.5.1 节）的 add 方法进行添加功能处理。添加成功跳转到查询视图；添加失败回到添加视图。

10.5.5 查询名片

管理员登录成功后，进入名片管理系统的主页面，在主页面中初始显示查询视图 queryCards.jsp，查询页面运行效果如图 10.6 所示。

单击主页面中"名片管理"菜单的"查询名片"菜单项，打开查询页面 queryCards.jsp。"查询名片"菜单项超链接的目标地址是个 url 请求。该请求路径为"card/query"，根据请求路径找到对应控制器类 CardController 的 query 方法处理查询功能。在该方法中，根据动作类型（"修改查询""查询"以及"删除查询"），将查询结果转发到不同视图。

名片ID	名称	单位	详情
6	张三	火星开发公司	详情
7	李四	月球开发公司	详情
8	王五	土星开发公司	详情
9	赵六	金星开发公司	详情
10	张四	水星开发公司	详情
11	李五	地球开发公司	详情
12	李六	牛郎开发公司	详情
13	王六	织女开发公司	详情
14	赵七	北斗星开发公司	详情
15	赵八	伽利略开发公司	详情
共11条记录 共2页 第1页 下一页			

图 10.6　查询页面

在 queryCards.jsp 页面中单击“详情”超链接，打开名片详细信息页面 detail.jsp。“详情”超链接的目标地址是个 url 请求。该请求路径为“card/selectACard”。根据请求路径找到对应控制器类 CardController 的 selectACard 方法处理查询一个名片功能。将查询结果转发给详细信息页面 detail.jsp。名片详细信息页面如图 10.7 所示。

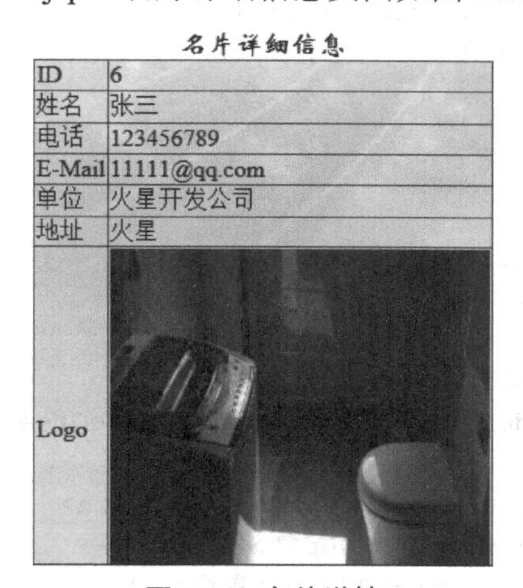

名片详细信息

ID	6
姓名	张三
电话	123456789
E-Mail	11111@qq.com
单位	火星开发公司
地址	火星
Logo	

图 10.7　名片详情

queryCards.jsp 的代码如下：

```
<%@ page language="java" contentType="text/html; charset=UTF-8"
pageEncoding="UTF-8"%>
<%@ taglib uri="http://java.sun.com/jsp/jstl/core" prefix="c" %>
<%
String path = request.getContextPath();
String basePath = request.getScheme()+"://"+request.getServerName()+":"+
request.getServerPort()+path+"/";
%>
<!DOCTYPE html PUBLIC "-//W3C//DTD HTML 4.01 Transitional//EN"
"http://www.w3.org/TR/html4/loose.dtd">
<html>
  <head>
    <base href="<%=basePath%>">
    <title>queryCards.jsp</title>
     <link href="css/common.css" type="text/css" rel="stylesheet">
```

```
    <style type="text/css">
        table{
            text-align: center;
            border-collapse: collapse;
        }
        .bgcolor{
            background-color: #F08080;
        }
    </style>
    <script type="text/javascript">
        function changeColor(obj){
            obj.className = "bgcolor";
        }
        function changeColor1(obj){
            obj.className = "";
        }
    </script>
</head>
<body>
    <c:if test="${allCards.size() == 0 }">
        您还没有客户。
    </c:if>
    <c:if test="${allCards.size() != 0 }">
        <table border="1" bordercolor="PaleGreen">
            <tr>
                <th width="200px">名片 ID</th>
                <th width="200px">名称</th>
                <th width="250px">单位</th>
                <th width="200px">详情</th>
            </tr>
            <c:forEach items="${allCards }" var="card">
                <tr onmousemove="changeColor(this)" onmouseout="changeColor1
(this)">
                    <!-- id、name 等都要与数据表的列名相同，因为 map 中的关键字
是它们 -->
                    <td>${card.id }</td>
                    <td>${card.name }</td>
                    <td>${card.company }</td>
                <td><a href="card/selectACard?id=${card.id }" target=
"_blank">详情</a></td>
                </tr>
            </c:forEach>
            <tr>
                <td colspan="4" align="right">
                       
           共${totalCount}条记录  共${totalPage}
```

```
页  
                    第${pageCur}页  
                    <c:url var="url_pre" value="card/query">
                        <c:param name="pageCur" value="${pageCur - 1 }"/>
                    </c:url>
                    <c:url var="url_next" value="card/query">
                        <c:param name="pageCur" value="${pageCur + 1 }"/>
                    </c:url>
                    <!-- 第一页没有上一页 -->
                    <c:if test="${pageCur != 1 }">
                      <a href="${url_pre}">上一页</a>   
 
                    </c:if>
                    <!-- 最后一页，没有下一页 -->
                    <c:if test="${pageCur != totalPage }">
                        <a href="${url_next}">下一页</a>
                    </c:if>
                </td>
            </tr>
        </table>
    </c:if>
</body>
</html>
```

detail.jsp 的代码如下：

```
<%@  page  language="java"  contentType="text/html;  charset=UTF-8"
pageEncoding="UTF-8"%>
<%@ taglib uri="http://java.sun.com/jsp/jstl/core" prefix="c" %>
<%
    String path = request.getContextPath();
    String basePath = request.getScheme() + "://" + request.getServerName()
+ ":" + request.getServerPort() + path + "/";
%>
<!DOCTYPE  html  PUBLIC  "-//W3C//DTD  HTML  4.01  Transitional//EN"
"http://www.w3.org/TR/html4/loose.dtd">
<html>
<head>
<base href="<%=basePath%>">
<title>detail.jsp</title>
</head>
<body>
    <center>
        <table border=1 background="images/bb.jpg" style="border-collapse:
collapse">
            <caption>
```

```
            <font size=4 face=华文新魏>名片详细信息</font>
        </caption>
        <tr>
            <td>ID</td>
            <td> ${acard.id }</td>
        </tr>
        <tr>
            <td>姓名</td>
            <td>${acard.name }</td>
        </tr>
        <tr>
            <td>电话</td>
            <td>${acard.telephone }</td>
        </tr>
        <tr>
            <td>E-Mail</td>
            <td>${acard.email }</td>
        </tr>
        <tr>
            <td>单位</td>
            <td>${acard.company }</td>
        </tr>
        <tr>
            <td>地址</td>
            <td>${acard.address }</td>
        </tr>
        <tr>
            <td>Logo</td>
            <td>
                <c:if test="${acard.logo != '' }">
                    <img alt="" width="250" height="250"
                    src="logos/${acard.logo}"/>
                </c:if>
            </td>
        </tr>
    </table>
  </center>
</body>
</html>
```

10.5.6 修改名片

单击主页面中“管理名片”菜单的“修改名片”菜单项，打开修改查询页面updateSelect.jsp。“修改名片”菜单项超链接的目标地址是个 url 请求。找到对应控制器类 CardController 的方法 query，在该方法中，根据动作类型，将查询结果转发给修改查询视图。

单击 updateSelect.jsp 页面中的“修改”超链接，打开修改名片信息页面 updateCard.jsp。“修改”超链接的目标地址是个 url 请求。找到对应控制器类 CardController 的方法 selectACard，在该方法中，根据动作类型，将查询结果转发给 updateCard.jsp 页面显示。

输入要修改的信息后，单击“提交”按钮，将名片信息提交给控制器类，找到对应控制器类 CardController 的方法 add，在 add 方法中根据动作类型，执行修改的业务处理。修改成功，进入查询名片。修改失败，回到 updateCard.jsp 页面。

updateSelect.jsp 页面的运行效果如图 10.8 所示，updateCard.jsp 页面的运行效果如图 10.9 所示。

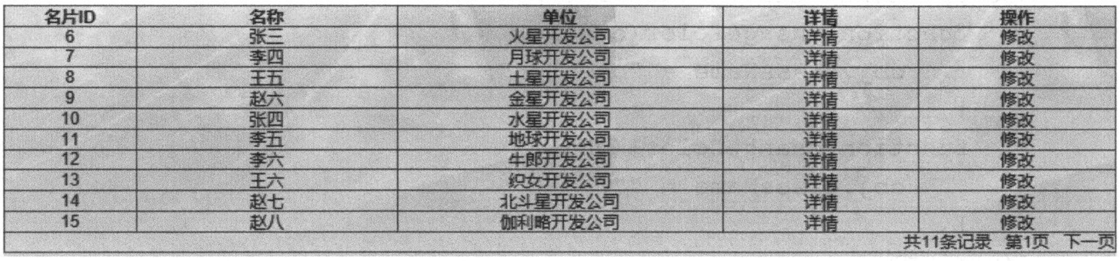

名片ID	名称	单位	详情	操作
6	张三	火星开发公司	详情	修改
7	李四	月球开发公司	详情	修改
8	王五	土星开发公司	详情	修改
9	赵六	金星开发公司	详情	修改
10	张四	水星开发公司	详情	修改
11	李五	地球开发公司	详情	修改
12	李六	牛郎开发公司	详情	修改
13	王六	织女开发公司	详情	修改
14	赵七	北斗星开发公司	详情	修改
15	赵八	伽利略开发公司	详情	修改
				共11条记录 第1页 下一页

图 10.8 updateSelect.jsp 页面

修改名片

ID*	6
名称*	张三
电话*	123456789
E-Mail	11111@qq.com
单位	火星开发公司
Logo	浏览...
提交	重置

图 10.9 updateCard.jsp 页面

updateSelect.jsp 的代码如下：

```
<%@ page language="java" contentType="text/html; charset=UTF-8"
pageEncoding="UTF-8"%>
<%@ taglib uri="http://java.sun.com/jsp/jstl/core" prefix="c" %>
<%
String path = request.getContextPath();
String basePath=request.getScheme()+"://"+request.getServerName()+":"+
request.getServerPort()+path+"/";
%>
<!DOCTYPE html PUBLIC "-//W3C//DTD HTML 4.01 Transitional//EN"
"http://www.w3.org/TR/html4/loose.dtd">
<html>
  <head>
```

```
        <base href="<%=basePath%>">
        <title>updateSelect.jsp</title>
         <link href="css/common.css" type="text/css" rel="stylesheet">
         <style type="text/css">
            table{
                text-align: center;
                border-collapse: collapse;
            }
            .bgcolor{
            background-color: #F08080;
         }
         </style>
         <script type="text/javascript">
            function changeColor(obj){
                obj.className = "bgcolor";
            }
            function changeColor1(obj){
                obj.className = "";
            }
         </script>
    </head>
    <body>
         <c:if test="${allCards.size() == 0 }">
            您还没有客户。
         </c:if>
         <c:if test="${allCards.size() != 0 }">
         <table border="1" bordercolor="PaleGreen">
            <tr>
                <th width="100px">名片 ID</th>
                <th width="200px">名称</th>
                <th width="250px">单位</th>
                <th width="150px">详情</th>
                <th width="150px">操作</th>
            </tr>
            <c:forEach items="${allCards }" var="card">
                <tr onmousemove="changeColor(this)" onmouseout="changeColor1
    (this)">
                    <!-- id、name 等都要与数据表的列名相同, 因为 map 中的关键字是它
    们 -->
                    <td>${card.id }</td>
                    <td>${card.name }</td>
                    <td>${card.company }</td>
                    <td><a href="card/selectACard?id=${card.id}" target=
    "_blank">详情</a></td>
         <td><a href="card/selectACard?id=${card.id }&act=updateAcard" target
    ="center">修改</a></td>
```

```
            </tr>
        </c:forEach>
        <tr>
            <td colspan="5" align="right">
               
               共${totalCount}条记录  
            第${pageCur}页  
            <c:url var="url_pre" value="card/query">
                    <c:param name="pageCur" value="${pageCur - 1 }"/>
                    <c:param name="act" value="updateSelect"/>
            </c:url>
            <c:url var="url_next" value="card/query">
                <c:param name="pageCur" value="${pageCur + 1 }"/>
                <c:param name="act" value="updateSelect"/>
            </c:url>
            <!-- 第一页没有上一页 -->
            <c:if test="${pageCur != 1 }">
                <a href="${url_pre}">上一页</a>    
            </c:if>
            <!-- 最后一页，没有下一页 -->
            <c:if test="${pageCur != totalPage }">
                <a href="${url_next}">下一页</a>
            </c:if>
            </td>
        </tr>
    </table>
    </c:if>
</body>
</html>
```

updateCard.jsp 的代码如下：

```
<%@ page language="java" import="java.util.*" pageEncoding="UTF-8"%>
<%@ taglib prefix="form" uri="http://www.springframework.org/tags/form"
%>
<%@ taglib uri="http://java.sun.com/jsp/jstl/core" prefix="c" %>
<%
String path = request.getContextPath();
String basePath = request.getScheme()+"://"+request.getServerName()+":"+
request.getServerPort()+path+"/";
%>
<!DOCTYPE HTML PUBLIC "-//W3C//DTD HTML 4.01 Transitional//EN">
<html>
  <head>
    <base href="<%=basePath%>">
    <title>My JSP 'updateCard.jsp' starting page</title>
```

```
    </head>
    <body>
    <form:form action="card/add?updateAct=update" method="post" model_
  Attribute="card" enctype="multipart/form-data">
            <table border=1 style="border-collapse: collapse">
                <caption>
                    <font size=4 face=华文新魏>修改名片</font>
                </caption>
                <tr>
                    <td>ID<font color="red">*</font></td>
                    <td>
                <form:input path="id" readonly="true" cssStyle="border-width:
                1pt; border-style: dashed; border-color: red"/>
                    </td>
                </tr>
                <tr>
                    <td>名称<font color="red">*</font></td>
                    <td>
                        <form:input path="name"/>
                    </td>
                </tr>
                <tr>
                    <td>电话<font color="red">*</font></td>
                    <td>
                        <form:input path="telephone"/>
                    </td>
                </tr>
                <tr>
                    <td>E-Mail</td>
                    <td>
                        <form:input path="email"/>
                    </td>
                </tr>
                <tr>
                    <td>单位</td>
                    <td>
                        <form:input path="company"/>
                    </td>
                </tr>
                <tr>
                    <td>地址</td>
                    <td>
                        <form:input path="address"/>
                    </td>
                </tr>
                <tr>
```

```
                    <td>Logo</td>
                    <td>
                        <input type="file" name="logo">
                        <br>
                        <!-- 从数据库取出的文件名 -->
                        <c:if test="${card.fileName != ''}">
                            <img alt="" width="50" height="50"
                            src="logos/${card.fileName}"/>
                        </c:if>
                    </td>
                </tr>
                <tr>
                    <td align="center"><input type="submit" value="提交"/>
</td>
                    <td align="left"><input type="reset" value="重置"/></td>
                </tr>
            </table>
        </form:form>
    </body>
</html>
```

10.5.7 删除名片

单击主页面中“管理名片”菜单的“删除名片”菜单项，打开删除查询页面 deleteSelect.jsp。“删除名片”菜单项超链接的目标地址是个 url 请求。根据请求找到对应控制器类 CardController 的方法 query，在该方法中，根据动作类型，将查询结果转发给 deleteSelect.jsp 页面，页面效果如图 10.10 所示。

ID	名称	单位	详情	操作
☐6	张三	火星开发公司	详情	删除
☐7	李四	月球开发公司	详情	删除
☐8	王五	土星开发公司	详情	删除
☐9	赵六	金星开发公司	详情	删除
☐10	张四	水星开发公司	详情	删除
☐11	李五	地球开发公司	详情	删除
☐12	李六	牛郎开发公司	详情	删除
☐13	王六	织女开发公司	详情	删除
☐14	赵七	北斗星开发公司	详情	删除
☐15	赵八	伽利略开发公司	详情	删除
		删除		
				共11条记录 第1页 下一页

图 10.10 deleteSelect.jsp 页面

在图 10.10 的复选框中选择要删除的名片，单击“删除”按钮，将要删除名片的 ID 提交给控制器类。找到对应控制器类 CardController 的方法 deleteCards，在该方法中，执行批量删除的业务处理。

单击图 10.10 中的“删除”超链接，将当前行的名片 ID 提交给控制器类，找到对应控制器类 CardController 的方法 deleteACard，在该方法中，执行单个删除的业务处理。

删除成功后，进入删除查询页面。

deleteSelect.jsp 的代码如下：

```
<%@  page  language="java"  contentType="text/html;  charset=UTF-8"
pageEncoding="UTF-8"%>
<%@ taglib uri="http://java.sun.com/jsp/jstl/core" prefix="c" %>
<%
String path = request.getContextPath();
String basePath = request.getScheme()+"://"+request.getServerName()+":"+
request.getServerPort()+path+"/";
%>
<!DOCTYPE  html  PUBLIC  "-//W3C//DTD  HTML  4.01  Transitional//EN"
"http://www.w3.org/TR/html4/loose.dtd">
<html>
  <head>
    <base href="<%=basePath%>">
    <title>deleteSelect.jsp</title>
     <link href="css/common.css" type="text/css" rel="stylesheet">
     <style type="text/css">
          table{
               text-align: center;
               border-collapse: collapse;
          }
          .bgcolor{
          background-color: #F08080;
     }
     </style>
     <script type="text/javascript">
          function confirmDelete(){
               var n = document.deleteForm.ids.length;
               var count = 0;//统计没有选中的个数
               for(var i = 0; i < n; i++){
                    if(!document.deleteForm.ids[i].checked){
                         count++;
                    }else{
                         break;
                    }
               }
               if(n > 1){//多个名片
                    //所有的名片都没有选择
                    if(count == n){
                         alert("请选择删除的名片！");
                         count = 0;
                         return false;
                    }
               }else{//一个名片
```

```
                //就一个名片并且还没有选择
                if(!document.deleteForm.ids.checked){
                        alert("请选择删除的名片！");
                        return false;
                }
            }

            if(window.confirm("真的删除吗？really?")){
                document.deleteForm.submit();
                return true;
            }
            return false;
        }
        function checkDel(id){
            if(window.confirm("是否删除该名片？")){
                window.location.href = "/cardsManage/card/ deleteACard?id="
+id;
            }
        }
        function changeColor(obj){
            obj.className = "bgcolor";
        }
        function changeColor1(obj){
            obj.className = "";
        }
  </script>
</head>
<body>
    <br>
    <c:if test="${allCards.size() == 0 }">
        您还没有客户。
    </c:if>
    <c:if test="${allCards.size() != 0 }">
    <form action="card/deleteCards" name="deleteForm">
    <table border="1" bordercolor="PaleGreen">
        <tr>
            <th width="250px">ID</th>
            <th width="200px">名称</th>
            <th width="200px">单位</th>
            <th width="200px">详情</th>
            <th width="200px">操作</th>
        </tr>
        <c:forEach items="${allCards }" var="card">
            <tr  onmousemove="changeColor(this)" onmouseout= "changeColor1
(this)">
                <td>
```

```
                    <input type="checkbox" name="ids" value="${card.id }"/>
                    ${card.id }
                </td>
                <!-- id、name 等都要与数据表的列名相同，因为 map 中的关键字是它
们 -->
                <td>${card.name }</td>
                <td>${card.company }</td>
                <td><a  href="card/selectACard?id=${card.id  }"  target=
"_blank"> 详情</a></td>
                <td>
                    <a href="javascript:checkDel('${card.id }')">删除</a>
                </td>
            </tr>
        </c:forEach>
        <tr>
            <td colspan="5">
                <input type="button" value="删除" onclick="confirmDelete ()">
            </td>
        </tr>

        <tr>
            <td colspan="5" align="right">
               
               共${totalCount}条记录  
            第${pageCur}页  
            <c:url var="url_pre" value="card/query">
                    <c:param name="pageCur" value="${pageCur - 1 }"/>
                    <c:param name="act" value="deleteSelect"/>
            </c:url>
            <c:url var="url_next" value="card/query">
                <c:param name="pageCur" value="${pageCur + 1 }"/>
                <c:param name="act" value="deleteSelect"/>
            </c:url>
            <!-- 第一页没有上一页 -->
            <c:if test="${pageCur != 1 }">
                <a href="${url_pre}">上一页</a>    
            </c:if>
            <!-- 最后一页，没有下一页 -->
            <c:if test="${pageCur != totalPage }">
                <a href="${url_next}">下一页</a>
            </c:if>
            </td>
        </tr>
    </table>
    </form>
```

```
    </c:if>
</body>
</html>
```

10.6 用户相关

与系统相关的 CSS 和图片位于 WebContent 目录下，与用户相关的 JSP 页面位于 WEB-INF/jsp/ user 目录下。

10.6.1 Controller 实现

UserController 控制器类负责处理“会员注册”“会员登录”“安全退出”以及“个人中心”的功能，具体代码如下：

```
package com.controller;
import javax.servlet.http.HttpSession;
import org.springframework.beans.factory.annotation.Autowired;
import org.springframework.stereotype.Controller;
import org.springframework.ui.Model;
import org.springframework.web.bind.annotation.ModelAttribute;
import org.springframework.web.bind.annotation.RequestMapping;
import com.domain.User;
import com.service.UserService;
@Controller
@RequestMapping("/user")
public class UserController {
    //依赖注入 UserService 进行后台处理
    @Autowired
    private UserService userService;
    /**
     * 登录
     */
    @RequestMapping("/login")
    public String login(@ModelAttribute User user, Model model, HttpSession
session) {
        if(userService.login(user)){
            session.setAttribute("user", user);
            return "/card/main";
        }
        model.addAttribute("errorMessage", "用户名或密码错误");
        return "/user/login";
    }
    /**
     * 注册
```

```
     */
    @RequestMapping("/register")
    public String register(@ModelAttribute User user, Model model) {
        if(user.getFlag() == 0){
            if(!userService.isExit(user))
                //用户名已存在
                model.addAttribute("isExit", "false");
            else
                //用户名可用
                model.addAttribute("isExit", "true");
            return "/user/register";
        }else{
            //注册成功
            if(userService.register(user))
                return "/user/login";
            //注册失败
            return "/user/register";
        }
    }
    /**
     * 安全退出
     */
    @RequestMapping("/exit")
    public String exit(HttpSession session, Model model){
        session.invalidate();
        model.addAttribute("user", new User());
        return "/user/login";
    }
    /**
     * 基本信息
     */
    @RequestMapping("/userInfo")
    public String userInfo(HttpSession session, Model model){
        User u = (User)session.getAttribute("user");
        model.addAttribute("user", u);
        return "/user/userInfo";
    }
    /**
     * 修改密码初始页面
     */
    @RequestMapping("/updatePWD")
    public String updateInput(HttpSession session, Model model){
        User u = (User)session.getAttribute("user");
        model.addAttribute("user", u);
        return "/user/updatePWD";
    }
```

```
    /**
     * 修改密码
     */
    @RequestMapping("/updateUser")
    public String updateUser(HttpSession session, String password, Model
model){
        User u = (User)session.getAttribute("user");
        model.addAttribute("user", u);
        if(userService.updateUser(u.getUsername(), password)){
            return "/user/login";
        }else{
            return "/user/updatePWD";
        }
    }
}
```

10.6.2 Service 实现

与用户管理相关的 Service 接口和实现类分别为 UserService 和 UserServiceImpl。

UserService 接口代码如下：

```
package com.service;
import com.domain.User;
public interface UserService {
    boolean login(User u);
    boolean register(User u);
    boolean isExit(User u);
    boolean updateUser(String username , String password);
}
```

UserServiceImpl 实现类代码如下：

```
package com.service;
import org.springframework.beans.factory.annotation.Autowired;
import org.springframework.stereotype.Service;
import com.dao.UserDao;
import com.domain.User;
//注解为一个服务
@Service
public class UserServiceImpl implements UserService{
    @Autowired
    private UserDao userDao;
    /**
     * 登录
     */
```

```
    @Override
    public boolean login(User u) {
        return userDao.login(u);
    }
    /**
     * 注册
     */
    @Override
    public boolean register(User u) {
        return userDao.register(u);
    }
    /**
     * 判断用户名是否存在
     */
    @Override
    public boolean isExit(User u) {
        return userDao.isExit(u);
    }
    /**
     * 修改密码
     */
    @Override
    public boolean updateUser(String username, String password) {
        return userDao.updateUser(username, password);
    }
}
```

10.6.3 Dao 实现

Dao 层是数据访问层，与用户管理相关的数据访问层为 UserDao，具体代码如下：

```
package com.dao;
import java.util.List;
import java.util.Map;
import org.springframework.stereotype.Repository;
import com.domain.User;
@Repository("userDao")
public class UserDao extends BaseDao{
    /**
     * 登录
     */
    public boolean login(User u){
        String sql = "select * from usertable where username=? and
password=?";
        Object obj[] = {
```

```
                    u.getUsername(),
                    u.getPassword()
                };
        List<Map<String, Object>> list = findByParam(sql, obj);
        if(list.size() > 0){
            return true;
        }else{
            return false;
        }
    }
    /**
     * 注册
     */
    public boolean register(User u){
        String sql = "insert into usertable values(?,?)";
        Object obj[] = {
                    u.getUsername(),
                    u.getPassword()
                };
        return updateByParam(sql,obj);
    }

    /**
     * 判断用户名是否存在
     */
    public boolean isExit(User u){
        String sql = "select * from usertable where userName=?";
        Object obj[] = {
                    u.getUsername()
                };
        List<Map<String, Object>> list = findByParam(sql, obj);
        if(list.size() > 0){
            return false;
        }else{
            return true;
        }
    }
    /**
     * 修改密码
     */
    public boolean updateUser(String username, String password){
        String sql = "update usertable set password=? where username=? ";
        Object obj[] = {
                password,
                username
            };
```

```
        return updateByParam(sql,obj);
    }
}
```

10.6.4 注册

在系统默认主页 index.jsp，单击“注册”链接，打开注册页面 register.jsp，效果如图 10.11 所示。

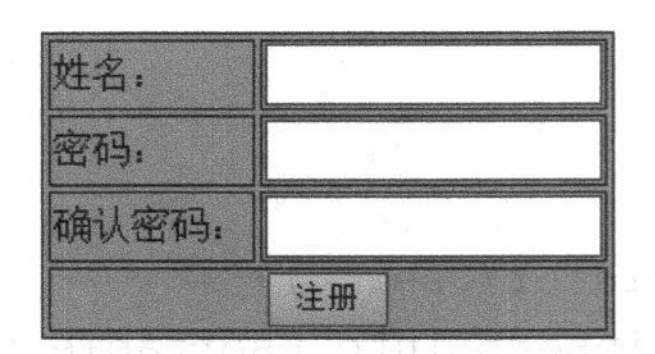

图 10.11 注册页面

在图 10.11 所示的注册页面中，输入“姓名”后，系统会根据请求路径“user/register”和标记位“flag”检测“姓名”是否可用。输入合法的用户信息后，单击“注册”按钮，实现注册功能。

register.jsp 的代码如下：

```
<%@ page language="java" contentType="text/html; charset=UTF-8"
pageEncoding="UTF-8"%>
<%@ taglib prefix="c" uri="http://java.sun.com/jsp/jstl/core" %>
<%@ taglib prefix="form" uri="http://www.springframework.org/tags/form"
%>
<%
String path = request.getContextPath();
String basePath = request.getScheme()+"://"+request.getServerName()+":"+
request.getServerPort()+path+"/";
%>
<!DOCTYPE html PUBLIC "-//W3C//DTD HTML 4.01 Transitional//EN"
"http://www.w3.org/TR/html4/loose.dtd">
<html>
<head>
<base href="<%=basePath%>">
<meta http-equiv="Content-Type" content="text/html; charset=UTF-8">
<style type="text/css">
    .textSize{
        width: 100pt;
        height: 15pt
    }
</style>
<title>注册画面</title>
<script type="text/javascript">
```

```
    //输入姓名后，调用该方法，判断用户名是否可用
    function nameIsNull(){
        var name = document.getElementById("username").value;
        if(name == ""){
            alert("请输入姓名！");
            document.getElementById("username").focus();
            return false;
        }
        document.getElementById("flag").value="0";
        document.registForm.submit();
        return true;
    }
    //注册时检查输入项
    function allIsNull(){
        var name = document.getElementById("username").value;
        var pwd = document.getElementById("password").value;
        var repwd = document.getElementById("repassword").value;
        if(name == ""){
            alert("请输入姓名！");
            document.getElementById("username").focus();
            return false;
        }
        if(pwd == ""){
            alert("请输入密码！");
            document.getElementById("password").focus();
            return false;
        }
        if(repwd == ""){
            alert("请输入确认密码！");
            document.getElementById("repassword").focus();
            return false;
        }
        if(pwd != repwd){
            alert("2 次密码不一致，请重新输入！");
            document.getElementById("password").value = "";
            document.getElementById("repassword").value = "";
            document.getElementById("password").focus();
            return false;
        }
        document.getElementById("flag").value = "1";
        document.registForm.submit();
        return true;
    }
</script>
</head>
<body>
```

```
    <form:form action="user/register" method="post" modelAttribute=
"user" name="registForm" >
        <form:hidden path="flag" id="flag"/>
        <table
        border=1
        bgcolor="lightblue"
        align="center">
            <tr>
                <td>姓名：</td>
                <td>
        <form:input id="username" path="username" cssClass="textSize"
onblur="nameIsNull()"/>
                    <c:if test="${isExit=='false' }">
                        <font color=red size=5>×</font>
                    </c:if>
                    <c:if test="${isExit=='true' }">
                        <font color=red size=5>√</font>
                    </c:if>
                </td>
            </tr>

            <tr>
                <td>密码：</td>
                <td>
        <form:password id="password" path="password" cssClass="textSize"
maxlength="20"/>
                </td>
            </tr>

            <tr>
                <td>确认密码：</td>
                <td>
        <form:password id="repassword" path="repassword" cssClass=
"textSize" maxlength="20"/>
                </td>
            </tr>

            <tr>
    <td colspan="2" align="center"><input type="button" value="注册"
onclick="allIsNull()"/></td>
            </tr>
        </table>
    </form:form>
</body>
</html>
```

10.6.5 登录

在系统默认主页 index.jsp，单击“登录”链接，打开登录页面 login.jsp，效果如图 10.12 所示。

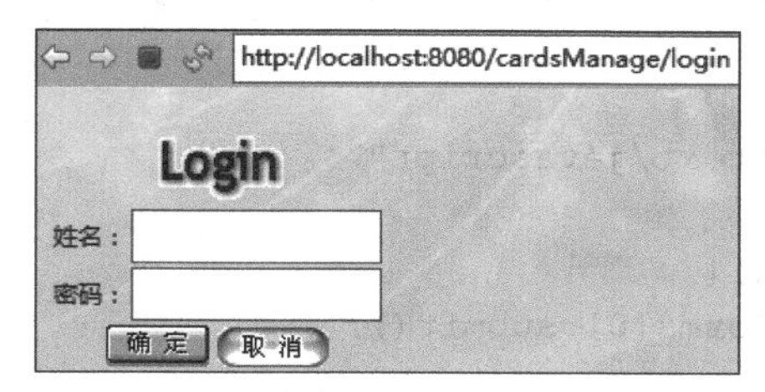

图 10.12 登录界面

用户输入姓名和密码后，系统将对姓名和密码进行验证。如果姓名和密码同时正确，则成功登录，将用户信息保存到 session 对象，并进入系统管理主页面（main.jsp）；如果姓名或密码有误，则提示错误。

login.jsp 的代码如下：

```
<%@ page language="java" contentType="text/html; charset=UTF-8"
pageEncoding="UTF-8"%>
<%@ taglib prefix="form" uri="http://www.springframework.org/tags/form" %>
<%
String path = request.getContextPath();
String basePath = request.getScheme()+"://"+request.getServerName()+":"+
request.getServerPort()+path+"/";
%>
<!DOCTYPE html PUBLIC "-//W3C//DTD HTML 4.01 Transitional//EN"
"http://www.w3.org/TR/html4/loose.dtd">
<html>
  <head>
  <base href="<%=basePath%>">
  <meta http-equiv="Content-Type" content="text/html; charset=UTF-8">
    <title>后台登录</title>
     <style type="text/css">
     table{
         text-align: center;
     }
     .textSize{
         width: 120px;
         height: 25px;
     }
     * {
         margin: 0px;
         padding: 0px;
```

```
    }
    body {
        font-family: Arial, Helvetica, sans-serif;
        font-size: 12px;
        margin: 10px 10px auto;
        background-image: url(images/bb.jpg);
    }
    </style>
    <script type="text/javascript">
    //"确定"按钮
    function gogo(){
        document.forms[0].submit();
    }
    //"取消"按钮
    function cancel(){
        document.forms[0].action = "";
    }
    </script>
  </head>
  <body>
    <form:form action="user/login" modelAttribute="user" method="post">
    <table>
        <tr>
            <td colspan="2"><img src="images/login.gif"></td>
        </tr>
        <tr>
            <td>姓名：</td>
            <td>
                <form:input path="username" cssClass="textSize"/>
            </td>
        </tr>
        <tr>
            <td>密码：</td>
            <td>
                <form:password path="password" cssClass="textSize" maxlength
="20"/>
            </td>
        </tr>
        <tr>
            <td colspan="2">
                <input type="image" src="images/ok.gif" onclick="gogo()" >
                <input  type="image"  src="images/cancel.gif"  onclick=
"cancel()" >
            </td>
        </tr>
    </table>
```

```
    </form:form>
    ${errorMessage }
  </body>
</html>
```

单击图 10.12 中的“确定”按钮，通过请求路径“user/login”，将登录请求提交给控制器类。根据请求路径找到对应控制器类 UserController（第 10.6.1 节）的 login 方法处理登录请求。

10.6.6 修改密码

单击主页面中“个人中心”菜单的“修改密码”菜单项，打开密码修改页面 updatePWD.jsp。页面效果如图 10.13 所示。

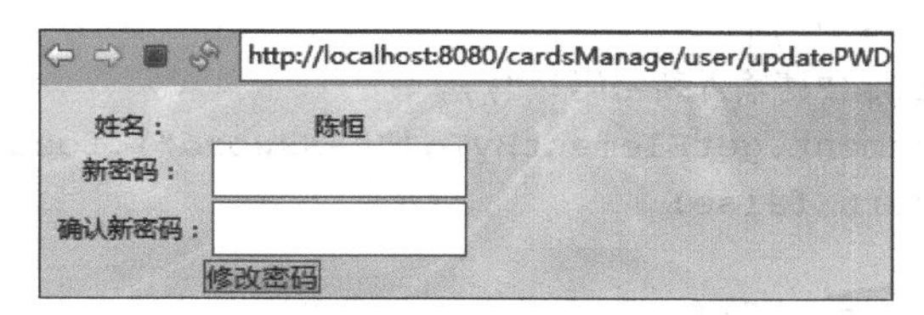

图 10.13　密码修改页面

updatePWD.jsp 的代码如下：

```
<%@ page language="java" import="java.util.*" pageEncoding="UTF-8"%>
<%@ taglib prefix="form" uri="http://www.springframework.org/tags/form" %>
<%
String path = request.getContextPath();
String basePath = request.getScheme()+"://"+request.getServerName()+":"+
request.getServerPort()+path+"/";
%>
<!DOCTYPE html PUBLIC "-//W3C//DTD HTML 4.01 Transitional//EN"
"http://www.w3.org/TR/html4/loose.dtd">
<html>
<head>
 <base href="<%=basePath%>">
<style type="text/css">
    table{
        text-align: center;
    }
    .textSize{
        width: 120px;
        height: 25px;
    }
    * {
        margin: 0px;
        padding: 0px;
```

```
        }
        body {
            font-family: Arial, Helvetica, sans-serif;
            font-size: 12px;
            margin: 10px 10px auto;
            background-image: url(images/bb.jpg);
        }
    </style>
    <title>修改密码</title>
    <script type="text/javascript">
        //注册时检查输入项
        function allIsNull(){
            var pwd = document.getElementById("password").value;
            var repwd = document.getElementById("repassword").value;
            if(pwd == ""){
                alert("请输入新密码！");
                document.getElementById("password").focus();
                return false;
            }
            if(repwd == ""){
                alert("请输入确认新密码！");
                document.getElementById("repassword").focus();
                return false;
            }
            if(pwd != repwd){
                alert("2 次密码不一致，请重新输入！");
                document.getElementById("password").value = "";
                document.getElementById("repassword").value = "";
                document.getElementById("password").focus();
                return false;
            }
            document.updateForm.submit();
            return true;
        }
    </script>
    </head>
    <body>
        <form:form action="user/updateUser" modelAttribute="user"  method=
    "post" name="updateForm">
            <table>
                <tr>
                    <td>姓名：</td>
                    <td>
                        ${user.username }
                    </td>
                </tr>
```

```
            <tr>
                <td>新密码：</td>
                <td>
                    <form:password id="password" path="password" cssClass
="textSize"/>
                </td>
            </tr>
            <tr>
                <td>确认新密码：</td>
                <td>
                    <form:password  id="repassword"  path="repassword"
cssClass="textSize"/>
                </td>
            </tr>
            <tr>
    <td colspan="2" align="center"><input type="button" value="修改密码"
onclick="allIsNull()"/></td>
            </tr>
        </table>
    </form:form>
</body>
</html>
```

在图 10.13 中输入“新密码”和“确认新密码”后，单击“修改密码”按钮，将请求通过“user/updateUser”提交给控制器类。根据请求路径找到对应控制器类 UserController（第 10.6.1 节）的 updateUser 方法处理密码修改请求。

10.6.7 基本信息

单击主页面中“个人中心”菜单的“基本信息”菜单项，打开基本信息页面 userInfo.jsp。页面效果如图 10.14 所示。

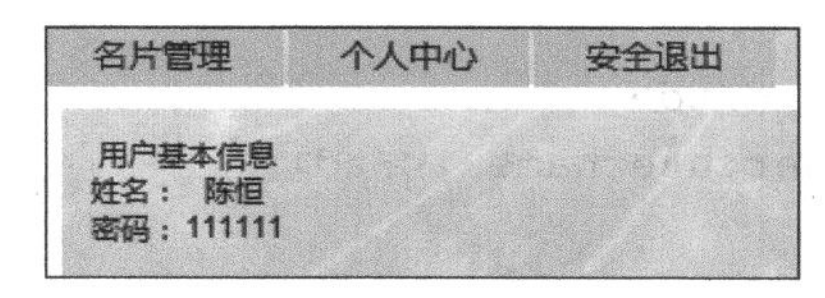

图 10.14 基本信息页面

userInfo.jsp 的代码如下：

```
<%@  page  language="java"  contentType="text/html;  charset=UTF-8"
pageEncoding ="UTF-8"%>
<%
String path = request.getContextPath();
String basePath=request.getScheme()+"://"+request.getServerName()+":"+
```

```
request.getServerPort()+path+"/";
%>
<!DOCTYPE html PUBLIC "-//W3C//DTD HTML 4.01 Transitional//EN"
"http://www.w3.org/TR/html4/loose.dtd">
<html>
  <head>
    <base href="<%=basePath%>">
   <title>用户基本信息</title>
    <style type="text/css">
    table{
        text-align: center;
    }
    .textSize{
        width: 120px;
        height: 25px;
    }
    * {
        margin: 0px;
        padding: 0px;
    }
    body {
        font-family: Arial, Helvetica, sans-serif;
        font-size: 12px;
        margin: 10px 10px auto;
        background-image: url(images/bb.jpg);
    }
    </style>
  </head>
  <body>
    <table>
        <tr>
            <td colspan="2">用户基本信息</td>
        </tr>
        <tr>
            <td>姓名：</td>
            <td>${user.username }</td>
        </tr>
        <tr>
            <td>密码：</td>
            <td>${user.password }</td>
        </tr>
    </table>
  </body>
</html>
```

10.7 安全退出

在管理主页面中，单击“安全退出”超链接，将返回后台登录页面。“安全退出”超链接的目标地址是一个请求处理方法，找到控制器类UserController（第10.6.1节）的对应处理方法exit。在该方法中执行：

```
session.invalidate();
```

将登录信息失效，并返回登录页面。

10.8 本章小结

本章讲述了名片管理系统的设计与实现。通过本章的学习，读者不仅掌握Spring MVC应用开发的流程、方法和技术，还应该熟悉名片管理的业务需求、设计以及实现。

参 考 文 献

[1] 史胜辉，王春明，陆培军．JavaEE 轻量级框架 Struts2+Spring+Hibernate 整合开发[M]．北京：清华大学出版社，2014．

[2] 范新灿．基于 Struts、Hibernate、Spring 架构的 Web 应用开发[M]．2 版. 北京：电子工业出版社，2014．

[3] 戴克．Spring MVC 学习指南[M]．林仪明，崔毅，译. 北京：人民邮电出版社，2015．

[4] 韩路彪．看透 Spring MVC：源代码分析与实践[M]．北京：机械工业出版社，2015．

[5] Budi Kurniawan，Paul Deck．Servlet JSP 和 Spring MVC 初学指南[M]．北京：人民邮电出版社，2016．